TECHNOLOGY TRANSFER

Technology Transfer

Edited by
DIETRICH SCHROEER AND MIRCO ELENA

Routledge
Taylor & Francis Group

LONDON AND NEW YORK

First published 2000 by Ashgate Publishing

Reissued 2018 by Routledge
2 Park Square, Milton Park, Abingdon, Oxon OX14 4RN
711 Third Avenue, New York, NY 10017, USA

Routledge is an imprint of the Taylor & Francis Group, an informa business

Publisher's Note
The publisher has gone to great lengths to ensure the quality of this reprint but points out that some imperfections in the original copies may be apparent.

Disclaimer
The publisher has made every effort to trace copyright holders and welcomes correspondence from those they have been unable to contact.

Typeset by Manton Typesetters, Louth, Lincolnshire, UK.

A Library of Congress record exists under LC control number: 99049273

ISBN 13: 978-1-138-70601-9 (hbk)
ISBN 13: 978-1-138-70600-2 (pbk)
ISBN 13: 978-1-315-20203-7 (ebk)

Contents

List of Tables

Preface

The chapters in this volume were originally presented as papers at the nineteenth summer course of the International School on Disarmament and Research on Conflicts (ISODARCO), held at Candriai (Trento), Italy, 25 August–3 September 1998.

The organization of the course was made possible by financial contributions from the University of Trento, University of Rome 'Tor Vergata', Istituto Trentino di Cultura, Regione Autonoma Trentino Alto Adige and the Fondazione Opera Campana dei Caduti di Rovereto.

We would like to express our deepest gratitude to several individuals without whom the course would not have been possible: Dr Mirco Elena and Prof. Maurizio Martellini, directors of the course, Prof. Dietrich Schroeer, and Dr Isabella Colace of the ISODARCO office in Rome.

For their hospitality, we are indebted to Ms Rita Malpaga and all the personnel of the Centro Attivita' Formative of the Provincia Autonoma di Trento at Candriai.

All opinions expressed in the individual chapters in this book are of a purely personal nature, and do not necessarily represent the official view of either the organizers of the course or of any organizations with which the writers may be affiliated.

Carlo Schaerf
President of ISODARCO and Director of the School

Introduction

The theme of the 1998 ISODARCO summer course was technology transfer, which has three major aspects:

1 the interchange of technologies between military and civilian applications – such topics as spin-off, dual use, conversion and diversification fall under this heading
2 the proliferation of weapons, either through arms races between developed nations, or through the transfer of military technology from developed arms industries to less-developed nations – including the topics of proliferation, arms races and arms control agreements
3 the transfer of civilian technologies from developing nations to less-developed nations – the expression of 'North–South transfer' and the idea of development fall under this heading.

The chapters in this book, originally presented as papers at the summer course, examine all three of these aspects.

PART I: TECHNOLOGY INTERCHANGE BETWEEN MILITARY AND CIVILIAN APPLICATIONS

Jordi Molas-Gallart (Chapter 1) explores the nature of dual-use technologies. He argues that technology involves not only equipment, but also the ability to use it – which for large technologies includes the infrastructure. Hence both the commercialization of military technologies and the militarization of commercial interests is not as easy as might be expected. Perhaps mixed industry may be most efficient, even if not necessarily most desirable.

Gary Chapman (Chapter 2) explores the impact the exponential growth of the technology of the Internet is having on national security, and vice versa. This technology is based on computers which have been the principal driving force in recent arms races. Examination of this issue reveals that this is an example of civilian technologies affecting the military sector. Much of the force driving the Internet's recent growth has come from the civilian sector, yet the Internet has become a critical asset for US national security. Recent attempts to re-impose military controls over this civilian technology have not been successful.

Ian S. Goudie (Chapter 3) examines a concrete example of the current efforts in many countries to civilianize military-based industries. The current conception is not so much defence conversion as diversification. He describes efforts to diversify a UK defence industry by trying to produce miliary and civilian versions of a laser-gyroscope simultaneously.

Dietrich Schroeer (Chapter 4) reviews the Global Positioning System (GPS), a technology that was transferred from the military to the civilian sector. Although the military developed this technology, civilian applications gradually overtook the military ones, to such an extent that the military is losing control over it. Is this a model for other conversion efforts?

Paolo Farinella, Luciano Anselmo and Bruno Bertotti (Chapter 5) describe nuclear power in space as one dual-use technology where the military use is threatening the civilian applications. Space junk from military satellite launchings endangers communications satellites, military nuclear reactors in space threaten scientific research capabilities, and space-based strategic defences would completely clutter space.

Dimitri Batani and Stefano Atzeni (Chapter 6) discuss the US and French nuclear stockpile stewardship programmes. These programmes are intended to allow the monitoring of the state of nuclear weapons without resorting to nuclear tests, which are banned under the Comprehensive Test Ban Treaty. The fact that these monitoring programmes also assist in designing nuclear weapons illustrates one of the major problems with dual-use technologies.

PART II: MILITARY TECHNOLOGY TRANSFER TO LESS-DEVELOPED COUNTRIES

Dingli Shen (Chapter 7) describes the efforts of China to prevent the transfer of nuclear weapons technologies to other nations. This commitment to nuclear non-proliferation is demonstrated by China's signing of the Nuclear Non-proliferation Treaty, and its participation in the Zangger Committee's controls on nuclear exports.

In contrast, Patricia Lewis (Chapter 8) questions how far export controls can actually reduce the proliferation of nuclear weapons. Instead, she suggests a greater emphasis on controlling nuclear weapons proliferation from the demand side.

Alexander DeVolpi (Chapter 9) evaluates the effectiveness of technological and procedural measures to prevent the smuggling of materials of national security concern. Can technology limit nuclear proliferation by controlling the transfer of relevant materials and technologies?

Jean Pascal Zanders (Chapter 10) examines the proliferation of chemical and biological weapons (CBWs). His interest lies in whether it is possible to reduce this CBW proliferation by controlling its demand side. The question is: what makes one country interested in CBWs, while another similar country rejects the CBW option?

Bruce D. Larkin (Chapter 11) examines the monitoring of agreements that limit either arms races or the transfer of military technologies. In particular, he asks how far less-developed countries and the public can monitor military affairs, particularly arms control and non-proliferation treaties.

PART III: TECHNOLOGY TRANSFER FOR DEVELOPMENT

Carlo Pietrobelli (Chapter 12) argues persuasively that social development is now inextricably tied to technology development, including transfer if indigenous technological capabilities are inadequate. Development aid to underdeveloped nations may have to focus on improvements in education and the build-up of supporting infrastructures.

Gert G. Harigel (Chapter 13) propose a high-tech solution to the development problems of developing nations. He argues that nuclear power may be a particularly suitable means of generating electricity for development.

Notes on the Contributors

Luciano Anselmo (Italy) is a research physicist at the National Research Council of Italy (CNR) in the Spaceflight Dynamics Section of the CNUCE Institute in Pisa. His research has focused on the effects of earth-orbiting space debris. He is co-author of *Detriti Spaziale: Un fattore di rischio che incombe sul futuro delle attivita' in orbita* ('Space Debris: A Risk Factor for the Future of Orbital Space Activities') (CUEN, Naples, 1999).

Stefano Atzeni (Italy) holds a PhD in nuclear engineering. He is a researcher at the Fusion Division of the Frascati Research Centre of ENEA (the Italian National Agency for Energy, New Technology and the Environment). His main scientific interests include the modelling of dense plasmas for inertial confinement fusion. His major publications include a review article on modelling of dense plasmas in *Plasma Physics and Controlled Fusion* (Vol. 29, 1987).

Dimitri Batani (Italy) is a researcher at the Physics Department of the University of Milan. He is working on the physics of inertial confinement fusion and on the use of laser-produced plasmas as high-brightness soft X-ray sources. His work on disarmament has focused on anti-missile defence systems, including the *Patriot* anti-tactical ballistic missile system.

Bruno Bertotti (Italy) is Professor of Astrophysics at the University of Pavia. His research interests include space physics, and he is participating in the Cassini space mission to the Saturnian system. He represents the Italian Space Agency in international organizations dealing with space debris. He is co-author of *Detriti Spaziale: Un fattore di rischio che incombe sul futuro delle attivita' in orbita* ('Space Debris: A Risk Factor for the Future of Orbital Space Activities') (CUEN, Naples, 1999).

Gary Chapman (USA) is Director of the 21st Century Project at the Lyndon B. Johnson School of Public Affairs of the University of Texas at Austin. He has published extensively in the area of computer-related technology issues, such as Internet-related security matters.

Alexander DeVolpi (USA) is a physicist at the Argonne National Laboratory in the USA. He has been principal investigator for projects involving arms control, verification, non-proliferation technology and policy for the US Departments of Energy and Defense. He has participated in the joint FAS/NRDC project on nuclear warhead dismantling, and has collaborated with colleagues in Europe and the former Soviet Union on various arms control demonstration projects. He has published *Born Secret: The H-Bomb, the Progressive Case, and National Security* (New York: Pergamon, 1981).

Paolo Farinella (Italy) is a researcher in the Department of Mathematics of the University of Pisa. He has written extensively on both the technical and social problems created by the increasing amounts of space debris in earth orbit.

Ian S. Goudie (UK) is Project Director of the Arms Conversion Project for the City of Glasgow. Before that, he worked in the UK defence industry, where he led the GEC-Ferranti Workers' Diversification Not Dole' campaign. He has written extensively on the restructuring of the UK defence industry, including co-authoring *The Aerospace Industry in Scotland* (Glasgow Arms Conversion Project, 1998) with Rob McNulty

Gert G. Harigel (Switzerland) is an Emeritus Senior Physicist at CERN. He is a Founding and Council Member of the International Network of Engineers and Scientists for Global Responsibility and Secretary/Treasurer of the Geneva International Peace Research Institute.

Bruce D. Larkin (USA) is Professor of Politics at the University of California at Santa Cruz. He is a specialist in arms control and Chinese foreign policy. He is author of *Nuclear Designs: Great Britain, France, and China in the Global Governance of Nuclear Arms* (Brunswick, NJ: Transactions, 1996).

Patricia Lewis (UK) is Director of United Nations Institute for Disarmament Research in Geneva. She was a consultant to the UK Government during the Conventional Forces in Europe Treaty negations, and has written extensively about the verification of arms control and disarmament treaties.

Jordi Molas-Gallart (Spain) is an economist and fellow at the Science Policy Research Unit of the University of Sussex, UK. He has written extensively on defence industrial policy, conversion and diversification strategies, and technology policy, including *Military Production and Innovation in Spain* (Newark, NJ: Gordon & Breach, 1992). Together with Julian Robinson, he recently conducted an assessment of dual technologies in the context of European security and defence for the European Parliament.

Carlo Pietrobelli (Italy) is Professor in Development Economics in the Department of Economy and Institutions of the University of Rome 'Tor Vergata'. He has written extensively about technology transfer and industrial partnerships for developing countries. He has published *Industry, Competitiveness and Technological Capabilities in Chile: A New Tiger from Latin America?* (London and New York: Macmillan and St Martin's Press, 1998).

Dietrich Schroeer (USA) is Professor of Physics at the University of North Carolina at Chapel Hill. He has taught courses on science and policy and on science, technology and military affairs for many years. He wrote the textbook *Science, Technology and the Nuclear Arms Race* (London: Wiley, 1984), and edited the 1996 ISODARCO Proceedings on *The Weapons Legacy of the Cold War* (Aldershot: Ashgate, 1997).

Dingli Shen (China) is a physicist by training, with post-doctoral studies on arms control at Princeton University. He is currently Director of the Office of International Programs and Deputy Director of the Center for American Studies at Fudan University in Shanghai. At Fudan University, he launched in 1991 and has since directed China's first university-based Program on Arms Control and Regional Security. He teaches course and conducts research on international relations, including acting as the chief editor of the yearbook *China Development Report*. He has helped to initiate the Shanghai Initiative, a four-nation dialogue on global and regional nuclear arms control and disarmament, involving high-level participants from China, India, Pakistan and the USA. In 1998, he organized the 6th Beijing-ISODARCO Seminar on Arms Control in Shanghai.

Jean Pascal Zanders (Belgium) is a research associate at the Swedish International Peace Research Institute. He is the SIPRI Chemical Weapons Project Leader. His publications include contributions to the SIPRI study on *The Challenge of Old Chemical Munitions and Toxic Armament Wastes* (Oxford: Oxford University Press, 1997), and to the *SIPRI Yearbook 1998: Armaments, Disarmament and International Security* (Oxford: Oxford University Press, 1998).

PART I
TECHNOLOGY INTERCHANGE
BETWEEN MILITARY AND
CIVILIAN APPLICATIONS

1   Dual-use Technologies and Transfer Mechanisms

Jordi Molas-Gallart

INTRODUCTION

The term 'technology transfer' has been applied to very different types of relationships. The focus of this chapter will be the transfer of dual-use technologies between military and civilian applications. It will stress the many different processes that are covered by this seemingly straightforward concept, and will examine the policy difficulties that the complexity of the issue generates.

This chapter will explore the notion of dual-use technology transfer by examining the meaning of each of its components in turn:

- What is 'technology'?
- What is 'dual-use'?
- Is the term 'dual-use technology' meaningful in itself, or is it just a useful concept that is used to justify specific industrial and technology control policies?

An overview of different types of dual-use technology transfer mechanisms follows. Finally, the chapter concludes with a brief discussion of policy implications.

DEFINITIONS OF DUAL-USE TECHNOLOGIES

Technology

It is tempting to launch into a discussion of technology transfer without stopping to consider our use of the term 'technology'. After all, we have an image of what technology entails: the product of human ingenuity as embodied in the many products that surround us. When we speak of 'high technology', we immediately think of supercomputers, clean rooms populated by men and women in white lab coats developing or manufacturing semiconductors, jet aircraft, ever-smaller mobile telephones, digital television, and so on. Our image of technology is dominated by products.

However, there is obviously more to technology than products. The means of producing products may also be considered 'technology'. In fact, academics have spent much time generating volumes analysing concepts of technology, and contrasting the relative merits of different definitions.[1] These discussions are important because they have the power to focus the attention of the users of such analyses (whether researchers, policy-makers, and so on) on specific aspects of the problems under investigation. For instance, if we understand 'technology' as artifacts and we discuss technology export controls, we may refer to lists of proscribed products and how these proscriptions are implemented. Talk about technology transfer then becomes a matter of how *items* used in one area of activity or in one place can be applied and used in other areas or places.

Some analysts have looked at 'technology' not as products, but as a broader concept encompassing the social relations and the mode of production in which the development and production of artifacts occur.[2] In this view of technology, 'technology transfer' means something very different: it refers to how modes of production have been moved from one society to another, and how they have changed the network of social relations characterizing the receiving communities.

These examples illustrate two opposite, extreme ways of viewing 'technology'. This chapter will adopt a concept of 'technology' half-way between understanding technology as artifacts and as whole systems of social relations. This concept is described by Antio and Laamanen in these terms: 'Technology comprises the ability to recognise technology problems, the ability to develop new concepts and tangible solutions to technical problems, the concepts and tangibles developed to solve technical problems, and the ability to exploit the concepts and tangibles in an effective way.'[3]

This definition provides a framework for studying the variety of skills, materials, artifacts and knowledge that can be applied to develop technical solutions that satisfy both military and civilian wants. Under this definition, hardware is still considered to be technology, since it represents tangible solutions to technical problems. Yet hardware alone is not enough to solve problems and satisfy the technological requirements of various sections of our societies. Technical problems must be properly recognized, solutions must be developed, and once such solutions have reached the product stage, the users will often require specialized knowledge and skills to exploit them efficiently. Knowledge and skills are embodied in people. If we are concerned about weapons proliferation, we might see 'technology controls' as a way of dealing with the problem. However, drafting lists of products that cannot be exported to certain countries and enforcing this prohibition addresses only part of the problem. If a country has acquired a new weapons system either as an 'off-the-shelf' system bought abroad or through the domestic assembly of (partly) imported components and sub-systems, scientists and technicians will usually have to participate in developing and deploying it efficiently. Therefore, labour mobility, in the form of technicians providing assistance to would-be proliferators, becomes a concern of technology controls: in fact, labour mobility becomes a form of technology transfer.

Similarly, when seeking technology transfer to exploit technological advances in new applications or geographical regions, it is not enough to look at the transmission and adaptation of products. For such a transfer, we also have to look at the changes the new potential producers and users must themselves undergo. Training issues become an integral component of technology transfer. In this way, the focus moves subtly from products to 'softer', people-related issues, such as how the new users could participate in the adaptation of a piece of equipment to meet their needs, and whether and how they should be trained. A different definition of 'technology' has provided us with a broader view of technology transfer.

Dual-use technologies

On the basis of the above definition of 'technology', it is now possible to analyse the specific kind of technology under consideration here: dual-use technologies. We define a technology as dual-use when it has both current or potential military and civilian applications.

Over the years, 'dual-use technologies' have been analysed from two different perspectives:

1 In the arms control literature, dual-use technologies present a *problem* when attempting to curb the international diffusion of weapons system.
2 Analysts of the relationship between military and civilian production have at times seen dual use as presenting an *opportunity* for the wider exploitation of research and manufacturing efforts beyond their initial (military *or* civilian) goals.

The first perspective was prevalent in the late 1970s and early 1980s. The 1977 United Nations report, *Economic and Social Consequence of the Arms*

Race and Military Expenditures,[4] already noted that when technologies were applicable to both military purposes and important civilian applications, attempts to control the arms race by restricting access to such technologies would inevitably conflict with attempts to support economic development by making these technologies available to all countries. This tension continues to be a source of disagreement and debate among those concerned with international peace and development issues.

More recently, the term 'dual-use technology' has often been employed in the second sense. It has been argued that the common technological base supporting both civilian and military technological development can, for instance, provide an opportunity for defence manufacturers to diversify into civilian operations, and/or to exploit commercial technologies for military applications.

However other analysts have objected to the use of the term in either of these senses, fearing that it may convey a misleading image, pitching military against civilian technologies as if they were clearly defined, contrasting entities, with dual-use technologies occupying a sort of middle ground between them. The reality is that there is a wide diversity of military products; some are exclusive to the military (such as 'stealth' ships), others are very different from equivalent civilian products (an attack helicopter may only share a few characteristics with a civilian transport helicopter), while others may be very similar or even identical (an example is the increased use by the military of commercial computers for specialized defence applications). There is also a common base of generic technologies which can be applied readily to either military or civilian developments. This is even more apparent if we adopt a broad view of technology. Once we consider skills as part of technology, the range of generic technologies with multiple applications becomes much broader. The organizational skills required to manage large projects, the skills to conduct research, to test equipment and to integrate complex systems can often be applicable to both military and civilian production.

Therefore, the distinction between military and civilian technologies is not clear-cut. Some have suggested that a more accurate image of these relationships would be achieved by referring to the multiple uses to which many technologies can be applied.[5] Some of them may be military, others civilian, but many technologies may well have multiple uses rather than just dual ones.

Another source of criticism of the 'dual-use' concept stems from the ways in which it has been used by some defence agencies to justify and develop policies to provide further support for traditional defence industries. For instance, in the USA the scope of dual-use policies has clearly moved from initial attempts at timid forms of conversion to policies designed to increase the efficiency of military production. The exploitation of commercial technologies for military applications is being pursued by many defence industries and their customers, and it is usually presented as a form of dual-use policy.

However, referring to 'dual use' does not automatically imply that the user views military and civilian technologies as clearly distinct. Nor does it preclude holding different policy positions. It is still reasonable to refer to 'dual-use technologies' as defined above. This focuses our attention on a specific set of problems: how can the efforts that have been invested in the development of weaponry and other military products now be harnessed for other, wealth-creating applications? Or conversely, depending on one's own policy agenda, how can the results of commercial R&D (research and development) investments be applied to the improvement be applied to the improvement of military products. We are therefore operating in a context in which we can clearly distinguish two separate sets of *applications*; talking about 'multiple uses' will not help our analysis here. Granted, these two separate sets of applications are not reproduced by two separate sets of technologies; hence it is important to note the diversity of dual-use technologies. The relationship between the military and civilian application of technological capabilities is very complex. In some cases, military and commercial applications will require similar product characteristics and manufacturing techniques; in others, the final requirements will be very different, and may demand that some parts of the processes will be handled by specialists with very different skills and in differing facilities. The way in which products, skills and facilities may move from one application to another are numerous, and vary from sector to sector. It is this variety which generates the diversity of dual-use technology transfer mechanisms, and that is the subject of the rest of this chapter.

WHAT IS DUAL-USE TECHNOLOGY TRANSFER?

When discussing technology transfer, we encounter a wide variety of forms of transfer. We are dealing with a very general concept that can be applied to the movement of technology across nations, across firms, between research and production stages, and across applications. Technology transfer occurs whenever technology (in one or more of its forms) is moved between economic units (within or across countries) *or* between applications. An analysis may focus on the effects of technology transfer from industrialized countries to less-developed economies, on transfers across firms within industrialized countries, on the ways in which technology moves between those that generate it in research facilities and those that apply it in production plants or in the supply of services, or on the transfer of technology from one application to another.

This chapter is only concerned with technology transfer across *applications*. Dual-use technology transfer is a special instance of technology transfer between applications that occurs when a dual-use technology that has been developed for a military (or civilian) use is transferred to a civilian (or military) application.

It is important to note that this type of transfer can take place without the technology being physically moved from one location to another. Most of the technology transfer literature concentrates instead on the problems that emerge when technologies do physically move across firms and countries. Yet we can move technology across applications within the confines of a plant: for instance, this will happen when firms engage in diversification or conversion processes. Employees will then try to develop new products and new skills, and the production facilities will have to be adapted; all these efforts will be part of a specific form of technology transfer. Therefore, the physical movement of technology is neither a necessary nor a sufficient condition for dual-use technology transfer to occur.

We can define two different forms of dual-use technology transfer, depending on whether the transfer takes place within the same unit (internal transfer) or across units (external transfer). For instance, a laboratory that has developed a component for use in a heat-seeking missile may try to adapt it for application to commercial surveillance equipment. The unit will deploy its technological capabilities to adapt a product that was initially developed for a military application to a civilian application. This process of adaptation occurs within the same unit, and is therefore a case of *internal* dual-use technology transfer. The same laboratory may decide instead to license the basic technology to a security firm, which will then be responsible for developing the new commercial product – this is a case of *external* transfer.

Finally, it must be noted that a dual-use technology can be transferred within and across units without any intention of changing its application. For example, the laboratory may sell a licence for a heat-seeking sensor to another company that intends to use the technology for the same military application. In this case, there has been a transfer across economic units, but not across purposes. There has been a transfer of technology, but it is not a case of 'dual-use technology transfer' as defined above. Dual-use technology transfer refers only to the case when there is an intention of changing the initial (military or civilian) application of a technology. Technology can, of course, flow in both directions.

Civilian to military

Exploiting technologies developed elsewhere for defence applications can help control the escalation in the costs of military products, and can provide a way of exploiting the technological dynamism of civilian production. As discussed above, defence procurement agencies across the world are recognizing these possibilities and are adapting their procurement policies accordingly. The different dual-use policies explicitly endorsed by the US, French and German defence ministries, among others, are designed with this in mind. Procurement policies are also becoming more flexible, providing mechanisms to allow commercial off-the-shelf components to be included within military systems whenever possible.

Military to civilian

There has long been a keen interest to exploit the fruits of defence R&D in civilian markets. Government expenditure on defence R&D is still very substantial: in the UK, for instance, it still represents over 40 per cent of total government R&D expenditure. Although much of this effort may have limited application outside the military field, defence R&D generates outputs and funds inputs (laboratories, equipment, skilled researchers) that may be exploited more broadly. The transfer of the results of defence R&D is often considered to be the main, if not the only, dual-use policy worth addressing. Many studies, like the UK Maddock Report, *Civil Exploitation of Defence Technology*, on the electronics sector, concentrate on this form of transfer.[6] However, technology is much more than the direct results of research, and dual-use technology transfer goes beyond the exploitation of tangible research outputs. The many different forms of technology that our definition encompasses translate into a variety of technology transfer mechanisms which can be stimulated by a large variety of policies – policies which sometimes may not appear to be directly related to the goals of technology transfer. Policies such as procurement reform and conversion and diversification strategies can all be considered to be dual-use policies, in that they affect the number and efficiency of dual-use technology transfer mechanisms. The variety of such mechanisms is the subject of the next section.

MODES OF DUAL-USE TECHNOLOGY TRANSFER[7]

We can distinguish between different forms of dual-use technology transfer depending on *what* kind of dual-use technologies are being transferred, *who* is involved (particularly whether the transfer takes place within or across economic units), and *how* the transfer occurs. The previous sections addressed the variety of dual-use technologies, and distinguished between transfer across and within units. In addition, we can distinguish two main models of dual-use technology transfer depending on whether the transfer mechanism concerns itself with adapting the technology for its new use or whether it aims only at a straightforward transfer, leaving further development and changes to the recipient of the technology:

1   Most transferred technologies will require changes and adaptations to be exploited successfully in their new environment. However, in the context of 'straight transfer' mechanisms, the responsibility for carrying out such adaptation falls entirely on the receiver of the technology, and is not the concern of the source of the technology nor the transfer mechanism.
2   'Adaptational-transfer' policies are those approaches to dual-use technology transfer that directly address the adaptation of the technologies to their new applications. In this case, the transfer mechanism includes carrying out all or part of the necessary adaptations.

The specific technology being transferred is not the only element that needs to be adapted. Often, the organizations involved will themselves need to change if the dual-use potential of the technologies at hand is to be fulfilled. Organizational change may involve the creation of new units, such as the British Dual-Use Technology Centres (see below), or changes in the way older units operate. A special case of organizational change occurs when previously discrete military and civilian activities are integrated. It has been common practice among military producers to keep their military and civilian operations strictly separate. Some industrial analysts have long argued that to benefit from economies of scale, and to exploit technological synergy between military and civilian development and production activities, old separations should be reduced or even eliminated. Military and civilian activities should then come together to share the same production facilities and personnel.[8] In such an integrated environment, the whole plant or research laboratory would be employed, without distinction, for both military- and civilian-related activities.

By combining the two modes of dual-use technology transfer (straight and adaptive) with the two main alternatives regarding the actors involved (whether the transfer is internal or external), we can define four main types of dual-use technology transfer mechanisms, as shown in Table 1.1.

These are fairly general categories, and within each one further distinctions could be made, depending, for instance, on the forms of technology

Table 1.1 The four main types of dual-use technology transfer mechanisms

Actors	Mode	
	No adaptation	*Adaptation*
Transfer internal to a single unit	Internal straight transfer	Internal adaptational transfer
Between two or more units	External straight transfer	External adaptational transfer

Source: Jordi Molas-Gallart, 'Which way to go? Defence technology and the diversity of "dual-use" technology transfer', *Research Policy*, Vol. 26 (1997), pp. 367–85.

being transferred (products fixed capital, labour). The following examples illustrate the variety of possible transfer mechanisms within each category.

Internal straight transfer

This includes mechanisms to move technologies across applications within the same business unit. Because the transfer mechanism itself is not concerned with the adaptation of the technology, 'internal straight transfer' does not aim at any structural change in the way the corporations operate. In particular, if military and civilian operations are separated, this separation will continue, and the division or section which receives the technology is entirely responsible for its further development and exploitation.

For example, the UK company GEC-Marconi established a policy to maintain a clear separation between its military and civilian activities.[9] Plant-based development laboratories are 'fed' from three central research centres that focus on developing enabling technologies of a more generic nature. In other words, these central research facilities develop a common pool of dual-use technologies to be transferred to the different civilian and military divisions. It is then up to the plant laboratories to develop the technologies further, engaging in a differentiation process for specific military and civilian applications.[10] Thus GEC-Marconi has put in place mechanisms to transfer technologies within the corporation, but leaving their adaptation to specific applications up to the receivers.

Another example of the internal straight transfer mechanism is the establishment of centralized product databases. This is a current trend in industrial sectors such as aerospace, responding to a wider strategy of applying sophisticated information technology tools to the design and engineering of complex systems. These databases hold data on all components and sub-systems designed by the corporation; such repositories of technical information can then be used for other designs. Therefore, even when military and civilian divisions are strictly separated, they contribute the output of their design work to the common data repositories, to which engineers from both civilian and military divisions have access.[11] In this way, the product data-base provides a channel for internal technology transfer.

The examples above relate to both components and generic technologies. End products can also be the object of internal straight transfer. The UK Ministry of Defence (MoD) has pursued a vigorous commercial off-the-shelf acquisition policy in the field of telecommunications.[12] Such policies provide an opportunity for commercial suppliers to obtain military business without having to modify their products. Such products are thus internally transferred, from being marketed only to civilian users to being marketed to military customers. Such internal straight transfer is contingent upon the producers pursuing new opportunities being opened in military markets.

Internal adaptational transfer

'Internal adaptational transfer' policies are concerned with the various processes of adaptation needed, so that internally transferred dual-use technologies can be exploited for their new intended applications. They will often be implemented in the context of wider corporate strategies addressing the changes in the balance between military and civilian markets. The following are only a few example of such policies: Electromagnetic Sciences, a US firm with some 800 employees which specialized in battlefield communications and missile guidance technology, adapted its technologies to develop warehouse management systems for inventory tracking;[13] in the 1970s, Kaman, a US supplier to the helicopter industry, applied its vibration technology to the production of acoustic guitars;[14] a subsidiary of the German firm Diehl, specializing in the production of rubber components for tracked vehicles such as tanks, developed a line of components for the automobile and electronics industries; and also in Germany, Atlas Elektronik used its defence radar technology to develop a low-frequency radar for obtaining images of underground structures.[15]

The internal adaptation of products in order to enter new markets is never an easy process. It usually takes several years of effort to develop new products and commercialize them in new markets. In a well-known case, it took the Rockwell company twenty years to bring to the market the microwave ovens developed from the radar microwave technology it had developed during the Second World War.[16]

External straight transfer

Instead of trying to exploit the potential of a dual-use technology through internal transfer, business units may want to sell the technology to other firms or laboratories, which will then be responsible for further development and commercialization. The mechanisms described in this section are not concerned with the adaptation of the technology, but only with facilitating or performing its transfer.

Usually, the transferring firm or laboratory will sell the rights to use the technology to a firm which is closer to the intended new markets. In this way, defence suppliers which have developed products for which a commercial application is envisaged do not have to run the risk of having to explore and enter new fields unfamiliar to them. Selling the technology constitutes a low-risk approach to the exploitation of the dual-use potential of new technologies. The exploitation of the technology in the new environment is left entirely up to the buyer, and there is no need for institutional change in the defence supplier. There are five different mechanisms to facilitate the meeting of potential sellers and buyers of dual-use technology, and the eventual conclusion of transactions.

Technology brokers

These are independent organizations which provide an interface between technology suppliers and potential customers. They operate as market facilitators for new technology. In the field of dual-use technologies, brokers have been used to identify technologies with civilian applications emerging from military research efforts, and to market them to commercial clients. Technology brokers can be independent, 'for-profit' companies set up specifically for this task, or government-supported organizations. An example was Defence Technologies Enterprise (DTE), a UK firm set up by financial and investment institutions with the support of the MoD in 1984 to transfer technologies generated in the MoD research establishments. A contract signed with the MoD allowed DTE personnel to be placed with various Defence Research Establishments to spot new developments of interest.[17] DTE also established a system of Associate Members, who paid a modest annual fee to be kept informed about innovations from the MoD research establishment with commercial potential, and by 1987 there were some 170 Associate Members. DTE rapidly built a large database of transferable technologies, and initially saw this offer of technologies as a measure of its own success. Yet, despite all the transfer possibilities that were apparently being disclosed through its growing database, it soon became clear that DTE was not being successful in concluding technology transfer contracts.[18] The company ceased to operate in 1990, and it has been pointed out that one of the reasons for this failure was that 'simply transferring blueprints or even examples of hardware will be insufficient because the missing tacit knowledge must be relearned before the technology can even be recreated at the new site ... Effective technology translation therefore requires careful handling of the tacit knowledge of its developers.'[19]

Establishing communication channels

Technology brokering involves actively scouting for technologies and marketing them to potential civilian customers. A more timid attempt to assist technology transfer between military and civilian uses is the provision of communication meetings between defence laboratories and defence firms on the one side, and commercial entrepreneurs on the other. The aim is to assist in the development of new markets for the technologies emerging from military-oriented research through the improvement of communication links between potential buyers and sellers of technology. It is hoped that market forces will do the rest and come up with optimal technology transfer mechanisms.

These initiatives are often sponsored by government agencies or industrial associations, and can take the form of workshops, seminars or conferences. US Technology Applications Reviews are an example of a structured effort to develop a market for technologies generated by defence facilities and programmes. They are based on meetings between researchers and industry to identify commercialization opportunities.[20]

Internally led commercialization

Firms and laboratories may launch their own non-mediated efforts to find customers for the technological outputs they generate. Firms actively involved in seeking customers for their technologies may organize their own internal technology-brokering offices, which will have the same role as an external broker. In the USA, for instance, defence laboratories have created their own technology transfer offices (ORTAs) responsible for systematically identifying transferable technologies, potential customers, and organizing all aspects of the transfer. In the UK, the newly created DERATEC agency may play a similar role for the main UK Defence Evaluation and Research Agency (DERA).

Although these examples are based on the establishment of formal offices and departments, the mechanisms to commercialize the technology may be more informal. The Royal Signals and Radar Establishment (RSRE), one of the MoD research establishments that was later amalgamated into the Defence Research Establishment, established its own ways to disseminate research outputs through firms with which long-term relationships had been established. The mechanism was thus supported by informal personal contacts, and was not channelled through formal, specialized offices.

User facilities

It is common for Defence Research Establishments to have facilities which can be of use to commercial companies. Wind tunnels and other test facilities can be 'rented out' to industry, thus exploiting their dual-use potential.

The use of facilities for alternative goals can go beyond the occasional renting-out, and can result in a fully-fledged conversion within the operation of the plant. These changes have occurred when a military research or production facility can no longer operate because of a lack of demand, but the defence customer is interested in retaining the facilities in case a need arises in the future. Military activities are abandoned, but care is taken to retain the facilities so that it is possible to revert to military production at very short notice whenever necessary. An example of this policy is the US ARMS programme, a US Army initiative which has provided $200 million to assist the contracting operators of government-owned plants to rent the facilities to commercial firms. The revenues generated help compensate for the decline in Army contracts, and prevent the closure of the facilities.

Services: Consulting and outsourcing

Consulting and outsourcing revolve around the provision of services, and can provide a mechanism for dual-use technology transfer when the labour (and fixed capital) used by the service supplier was previously employed in other activities. For instance, military producers are increasingly outsourcing R&D tasks to smaller specialized companies; these firms are rarely specialized defence suppliers and operate in a dual-use environment. The result is a widening of the defence–civilian technology transfer channels, with the suppliers applying to their new defence work the knowledge, facilities and experience gained in their commercial operations.

The reverse can also happen: Defence Research Establishments are increasingly offering consulting services to civilian customers. The Defence Evaluation and Research Agency, for instance, has been actively marketing consulting services. As a result of these efforts, DERA has been contracted by a software entertainment company to develop audio templates for sound representation, and has developed new acoustic materials under contract from the medical industry.[21] Here, DERA is transferring the skills of its researchers and its laboratory facilities.

External adaptational transfer

'External adaptational transfer' occurs when the transfer mechanisms aim to move technology between partners and across applications, and simultaneously adapt the technology to the new requirements. What is being transferred could be the outputs of R&D efforts in the form of products, process technologies, new materials, and so on. The process of adaptation will often involve personnel from the generating facility (firm or laboratory) working together with the prospective user. As we will see from the following examples, this form of transfer is likely to generate new organizations to manage the new partnerships. In other words, external adaptational mechanisms typically involve organizational changes.

Spin-off companies

As has already been pointed out, many defence companies and laboratories are aware of the dual-use potential of some of the technologies they use or generate, but they feel unable to develop and commercialize the technologies in unfamiliar markets. Selling the technology to commercial firms is one of the alternatives addressed above. Another possibility is to create new firms (spin-off companies) to develop and commercialize ideas and technologies. This is a common mechanism for the commercialization of innovations emerging from government research laboratories. In the USA, more than fifty companies spun off from the federal laboratory system over a ten-year period.[22] These firms are usually created by former employees of the laboratories to exploit technologies and ideas they developed in their research.

Collaborative partnerships

Instead of approaching technology transfer as a one-off commercial transaction, the transferer and recipient may choose to work together to adapt the technology to its new applications. Because of the difficulties of such adaptation, collaborative partnerships require a deep commitment from the firm generating the technology that is to be transferred. The transferer must invest time and effort in the process, and both partners have to work out the details of the collaboration, from legal issues to practical matters such as management responsibilities and work location.[23] Most collaborative partnerships involve cost-sharing between the partners, but at times public support has been offered to set up collaborative ventures, particularly when they involve partners from different areas of technological activity, such as research laboratories, universities and companies. There was various ways in which such partnerships can be organized.

Publicly supported co-operative R&D programmes   Co-operative R&D programmes aim to improve the interface between industry, government and private and academic research facilities by establishing mechanisms for research collaboration. Such co-operative research and development is often prompted by government programmes providing financial incentives and creating mechanisms to conduct joint research projects. Although they are rarely driven by defence agencies, and therefore may not have a specific dual-use character, they often involve military-related facilities (laboratories or firms) in dual-use research. An example is the US Cooperative Research And Development Agreements (CRADAs). CRADAs provide a mechanism allowing government-operated laboratories, most of them heavily involved in defence work, to enter into research agreements with non-government parties. They have provided a channel for the commercial application of the results of defence-related research.

In the 1980s, the UK Department of Trade and Industry launched its Co-located Research Initiative Scheme. This policy aimed to stimulate collaborative research through the placement of staff from companies in the laboratories of the Defence Research Establishments. One example was the National Electronics Research Initiative (NERI). Typically, a NERI project would have about ten collaborating organizations, which contributed one researcher each, to be based at the host site in a defence laboratory.[24]

Joint centres   Defence and civilian laboratories and firms may come together to set up specific research centres for work on the development of dual-use technologies. An example of such an attempt is the Dual Use Technology Centres (DUTCs) being set up in the UK. Although DUTCs have been implemented in many different ways, their common objective is to facilitate the exploitation of defence-driven research for civil and commercial purposes: 'The concept of DUTCs is that the research in the DRA [Defence Research Agency] should recognise the potential for dual use from the outset, and allow industrial, academic or other government departments with similar technology but different market requirements to join in as partners.'[25]

Direct finance of dual-use R&D   Government agencies, particularly in the USA, have established research programmes to promote dual-use research involving partnerships between defence firms and laboratories on the one hand, and commercial enterprises on the other. One example was the now ailing US Technology Reinvestment Program (TRP), an ambitious initiative to create technology partnerships to develop dual-use products on cost-sharing basis, involving collaboration between defence prime suppliers and commercial firms.[26]

DUAL-USE TECHNOLOGY TRANSFER AS A DOUBLE-EDGED SWORD

This chapter has stressed the diversity of dual-use technology transfer mechanisms. The relationship between military and civilian technologies and production are complex, therefore we should not be tempted to generalize when describing the value of dual-use policies.

Significantly, this diversity also applies to the goals of dual-use technology transfer. Once channels for the transfer of technology across applications have been established, technologies can move from military to civilian uses, and equally from civilian to military applications. The existence of a shared common technological base between areas of military and civilian production, and of elements of technological convergence between military and civilian requirements, can help defence firms to move into civilian markets, as well as civilian companies to move into defence activities. Therefore, dual-use policies promoting the establishment of dual-use technology transfer mechanisms can not only be used to assist in the diversification (or even conversion) of defence-related facilities, but also to support military production, making it more efficient in a period of budgetary constraints. Further, if defence firms are able to diversify and penetrate new markets, they are bound to become more efficient in military production. In short, the potential for dual-use technologies can be exploited both ways.

It is far from clear how to interpret this duality from a policy point of view. On the one hand, some analysts are concerned about the implications for weapons proliferation arising from a more integrated defence and civilian industrial base. It has been argued that because of the widespread access to commercial technology in the context of globalization, the emphasis on commercial technology as a source of military strength is particularly dangerous and unsustainable in the long run. From this point of view, maintaining a separate military technology base will allow a country 'to capitalise on its distinctiveness in order to inhibit proliferation'.[27] Besides, an integrated civil/military industry and the use of common components and sub-systems in both military and civilian production can make it more difficult to track and monitor the development of military capabilities and may therefore pose a problem to anybody concerned with proliferation matters. These arguments seem to suggest that dual-use policies should be avoided for the military sector.

On the other hand, the existence of scientific and industrial enclaves that are wholly dependent for their livelihood on military production have traditionally supplied a powerful force for the maintenance of defence budgets and military production. Also, firms that need to export arms to maintain their economic viability will go to great lengths to seek foreign markets, even if this requires the circumvention of international and national legislation. To minimize these problems, it could be argued that flexible defence producers, capable of penetrating and operating in commercial markets, will be able to adapt more easily to reductions in military procurement budgets. As the balance between military and civilian markets favours the latter, flexible companies will become less dependent on military activities. When military production is only a small part of the overall activity of companies, they are less likely to risk the consequences of breaking or circumventing technology control and arms control legislation. If they are found to be acting in an irresponsible way in their military-related operations, the negative implications for their images will far outstrip the direct economic benefits gained by 'smuggling out' sensitive technologies. It has been pointed out that the collaboration of the chemical industry in setting up a chemical weapons control regime proves this point. For large chemical companies, the profits that could be obtained by getting involved in chemical weapons programmes are marginal; their own interests therefore lie in enforcing such a control regime. If chemical firms had been dependent on military-related trade, their co-operation would have been more difficult to obtain.[28] The argument continues that proliferation can therefore be more easily controlled in a framework of diversified firms involved in both civil and military production. From this point of view, dual-use policies deserve support.

There is a widespread belief in defence industrial sectors, and among policy analysts concerned with defence and proliferation issues, that the notion of 'dual use' is now going out of fashion. This could well be true. Some of the dual-use policies that received much attention in the USA have not been able to deliver all that was promised. And for reasons discussed above, groups concerned about weapons proliferation have been sceptical for some time about the value of the notion and the policies that are being introduced under the banner of 'dual use'. However, the policy dilemma presented here will remain whatever concepts we use to develop, present and justify defence industrial policies. The relationship between military and civilian production is in a state of flux, and the diversity of transfer channels described in this chapter will continue to apply for many years to come.

ACKNOWLEDGEMENT

The author wishes to acknowledge the support of ESRC Grant No. R000236926.

REFERENCES

1   See, for example, Kevin Willoughby, *Technology Choice: A Critique of the Appropriate Technology Movement* (Boulder, CO, and San Francisco, CA: Westview Press, 1990), pp.15–43.
2   Johan Galtung, *Development, Environment and Technology: Towards a Technology for Self-reliance*, United Nations Conference on Trade and Development, Report No. TD/B/C.6/23/Rev. 1 (New York: UNCTAD, 1979).
3   Errko Autio and Tomi Laamanen, 'Measurement and evaluation of technology transfer: Review of technology transfer mechanisms and indicators', *International Journal of Technology Management*, Vol. 10, Nos. 7/8 (1995), pp.643–64; p.647.
4   Secretary-General of the United Nations, *Economic and Social Consequences of the Arms Race and its Extremely Harmful Effect on World Peace and Security*, report to the 33rd Session of the General Assembly, A/32/88 (New York: UN, 1997), para. 146.
5   John A. Alic, Lewis M. Branscomb, Harvey Brooks, Ashton B. Carter and Gerard L. Epstein, *Beyond Spinoff: Military and Commercial Technologies in a Changing World* (Boston: Harvard Business School Press, 1992).
6   Ieuan Maddock, *Civil Exploitation of Defence Technology and Observations by the Ministry of Defence* (London: Economic Development Committee, February 1983).
7   The different types of dual-use technology transfer mechanisms addressed in this section of the chapter are discussed more fully in a previous article by the author: Jordi Molas-Gallart, 'Which way to go? Defence technology and the diversity of "dual use" technology transfer', *Research Policy*, Vol. 26 (1997), pp.367–85.
8   Jacques S. Gansler, 'Restructuring the Defense Industrial Base', *Issues in Science and Technology*, Vol. 8, No. 3 (1992), pp.49–58.
9   The separation is justified on the grounds of the strong differences between military and civilian production.
10 Daniel S. Gruneberg, 'The defence firm and trends in civil and military technologies: Integration versus "diversification"', in Andrew Latham and Nicholas Hooper (eds), *The Future of the Defence Firm: New Challenges, New Directions*, NATO ASI Series, Vol. 79 (Dordrecht: Kluwer Academic Publishers, 1995), pp.97–101.
11 An example of this approach to technical data management can be found in the company Rolls-Royce plc.
12 David Miller, 'Military communication goes civil', *International Defence Review*, No. 11 (1995), pp.56–7.
13 William Smith, 'How a prescient Pentagon contractor entered civilian life', *New York Times*, 12 April 1992, p.F–9.
14 A. Markusen and J. Yudken, *Dismantling the Cold War Economy* (New York: Basic

Books, 1992), p.213. This effort received a lot of publicity, but affected only 100 workers – Kenneth L. Adelman and Norman R. Augustine, 'Defense conversion: Bulldozing the management', *Foreign Affairs*, Vol. 71, No. 2 (Spring 1992), pp.26–47.

15 A. Zaks, *Diversification et reconversion de l'industrie d'armament* (Brussels: GRIP, 1992).

16 Alic, et al., *Beyond Spinoff*.

17 Bernard Herdan, 'A UK initiative for the transfer of technologies from defence to civil sector', in Philip Gummet and Judith Reppy (eds), *The Relations between Defence and Civil Technologies* (Dordrecht: Kluwer Academic Publishers, 1988), pp.159–65.

18 By the spring of 1987, only nine licence agreements had been concluded.

19 Graham Spinardi, 'Defence Technology Enterprises', *Science and Public Policy*, Vol. 19, No. 4 (1992), pp.198–206; p.205.

20 Technology assessment reviews were first developed for the Ballistic Missile Defense Organization (BMDO); T.R. Tucker, L.E. Aitcheson, and J.W. Reynolds, 'The Technology Applications Review: A catalyst for commercialization of defense technologies', paper presented at the 1994 Annual Meeting of the Technology Transfer Society (Huntsville, AL, 1994).

21 James Kerr and S. Birley, *A Report for the DTI on the Development of a Methodology for the Transfer of Generic Defence Technology to the Civil Sector* (Glasgow: National Engineering Laboratory, Report No. DUTA01, 25 May 1995).

22 Robert K. Carr, 'Doing technology transfer in federal laboratories', in Suleiman Kassicieh and Raymond Radosevich (eds), *From Lab to Market: Commercialization of Public Sector Technology* (New York and London: Plenum Press, 1994), pp.61–87.

23 Because of this need for substantial commitment, defence research laboratories may find it difficult to engage in these transfer mechanisms when technology transfer is not part of their mission. See Carr, 'Doing Technology Transfer'.

24 Adrian L. Mears, 'Technology transfer – Current and future methods', paper presented at the CERN Spring Conference (Geneva, 23 April 1987).

25 Office of Science and Technology, *Technology Foresight: Progress Through Partnership, Vol. 12: Defence and Aerospace* (London: HMSO, 1995), p.106.

26 US Department of Defense, *Dual Use Technology: A Defense Strategy for Affordable, Leading-edge Technology* (Washington, DC, US Government Printing Office, 1995). Although the initial focus was on the commercial exploitation of defence technologies, the policy later shifted towards the military exploitation of technologies generated elsewhere.

27 Judith Reppy, 'Dual use technology: Back to the future?', in Ann Markusen and Sean Costigan (eds), *Arming the Future: A Defence Industry for the 21st Century* (Washington, DC: Council for Foreign Relations, 1999).

28 Jacques Gansler, *Defense Conversion: Transforming the Arsenal of Democracy* (Cambridge, MA: MIT Press, 1995).

2 National Security and the Internet

Gary Chapman

INTRODUCTION

The modern concept of 'national security' and the electronic digital computer are roughly the same age, and both are products of the Second World War. ENIAC, the world's first digital electronic computer, went into service at the University of Pennsylvania in 1946. The US Government's Central Intelligence Agency and National Security Agency were launched a year later, authorized by the National Security Act.

Until relatively recently, national security and computers enjoyed a symbiotic relationship. Until the mid-1960s, perhaps even later, the chief US Government agencies responsible for national security were the chief catalysts and funders for computer research, and also the largest customers of the computer industry. Indeed, the appearance of the digital computer even shaped the strategy of national security in the USA, as more and more national security planning became dependent on computer-based models using techniques of systems analysis and operations research. One might even argue that this symbiotic relationship between computers and national security is the primary bearer and symbol of US power in the latter half of the twentieth century, even more so than nuclear weapons.

Computer technology is still important to national security – perhaps of paramount importance. Without computers, modern arsenals and 'battle management' and communications would be impossible. The future appears to belong to so-called 'smart' weapons, complex systems of command and control, telecommunications, satellites, electronic surveillance, and split-second information-processing. The end of the Cold War seems to have accelerated the process of integrating advanced computers into weapons and command systems, rather than to slowing it down. The USA's

23

overwhelming superiority in information technologies is the key to its superpower status for the foreseeable future.

But a new phenomenon is the threat to national security posed by networked computers, particularly through the Internet, as a result at the increasing use of the Internet for civilian purposes. The Internet is perhaps the most impressive example of technology transfer from the military sector to the civilian sector ever, so the threat is accompanied by more than a small amount of irony, as the Internet was, for decades, a project of the US Department of Defense (DoD). For a long time, during the period when the Internet was used almost exclusively by scientists, engineers, academics and a handful of military personnel, the Internet was viewed by experts mainly as a benign and interesting research project, with modest and limited applications to national security objectives. However, by the end of the 1990s, the Internet was increasingly regarded by national security officials as a new playing field for international conflict, a new medium in which national security will take on new forms, and one in which the US Government agencies responsible for national security have a growing stake. High officials of the Central Intelligence Agency, the National Security Agency, the Federal Bureau of Investigation, the White House, and other less well-known agencies now believe that the Internet is a 'critical national asset' that requires their attention and protection. This may signal a new era in the development of the Internet, equal in importance to its commercial potential. In fact, the commercial use of the Internet may be influenced by national security controversies as much as by consumer response to new Internet applications.

This chapter will review this controversy, looking first at the history of the Internet's relationship to national security, then providing an overview of the new landscape now that the Internet is increasingly embedded in 'critical national infrastructure', and yet at the same time is an increasing component of the civilian industrial sector. The concept of 'infowar', or 'cyberwar', will be described, along with the attendant difficulties of assessing computer-based threats to national assets. Finally, the chapter will offer some thoughts on what this new phenomenon might mean for the future development of the Internet, what strategies policy-makers and technology experts should consider, and what dangers lie ahead for democracy and public policy.

THE INTERNET AND THE MILITARY IN HISTORICAL CONTEXT

The Internet was first launched as a result project funded and managed by the US DoD Advanced Research Projects Agency (ARPA) in the late 1960s.

In 1983, the Defense Communications Agency split the network into two parts, ARPANET and MILNET, the former for the research community, and the latter for non-classified military communications. ARPANET's name was changed to the Internet, and its management was turned over to the National Science Foundation. It was also in 1983 that the network adopted TCP/IP (Transmission Control Protocol/Internet Protocol), perhaps the most important technical decision in the history of the Internet to date, allowing a vast expansion of the Internet that continues at an amazing rate of growth today.

There is a persistent myth surrounding the history of the Internet that it was designed to 'sustain a nuclear attack', and that this was the chief research interest of the Internet's Pentagon sponsors. As described in the definitive history of the Internet, *When Wizards Stay Up Late*, by Katie Hafner and Matt Lyon, the story that lies behind this myth is somewhat complicated.[1]

Hafner and Lyon write that Paul Baran, who joined the Air Force-sponsored thinktank RAND in Santa Monica, California, in 1959, 'developed an interest in the survivability of communications systems under nuclear attack': 'He was motivated primarily by the hovering tensions of the cold war, not the engineering challenges involved … Baran knew … that the early command and control systems for missile launch were dangerously fragile'.[2] At this time, during the late 1950s and early 1960s, the RAND Corporation was the primary source of strategic thinking for US nuclear policy, and the institution was already heavily dependent on computer technology, producing many of the earliest computer models of nuclear war.

RAND researchers had been working, without much success, on sustainable communications systems before Baran joined their ranks. It was Baran's theoretical work on distributed networked systems that pointed towards a solution. Baran came up with three theoretical innovations that became fundamental to the development of the Internet: a distributed network, network redundancy, and message disaggregation. This was a radical departure from the then universal model of communications based on centralized switching and open, direct circuits.

Baran's work was understood by only a handful of communications experts in the USA, and it was poorly received by the people in charge of improving defence communications, most of whom came from careers rooted in the more conventional model. He halted his work in 1964, convinced that the agencies responsible for military communications would botch the job even if they adopted his ideas: 'So I told my friends in the Pentagon to abort this entire program – because they wouldn't get it right,' he told Hafner and Lyon.[3] Instead, he decided to wait for the right moment, with some different kind of organization.

His opportunity emerged a few years later, when Larry Roberts, one of the ARPA officials in charge of investigating computer networks in the late 1960s, discovered Baran's RAND papers. However, Hafner and Lyon note that 'Nuclear war scenarios, and command and control issues, weren't high on Roberts' agenda'.[4] Roberts was intrigued by Baran's theoretical ideas of a distributed network from a purely research point of view. Roberts was also interested in a network that would tie together several of ARPA's chief research sites, universities and other institutions conducting experiments funded by the agency. It was Roberts who laid the first foundations of the Internet, relying on contributions from many different sources, including Baran, who became a consultant to the project. Thus, while Baran's work was motivated by the goal of building a communications network that could survive a nuclear war, this motivation was only a small part of the flow of ideas that built the technical foundations of the Internet.

Even more important is the fact that the Internet was never linked to any critical military application or system. For example, the Internet never played a role in controlling nuclear weapons. The communications network that connected US nuclear facilities, such as between the North American Air Defense Command in Cheyenne Mountain, Colorado – the hub of the country's 'early warning' system – and the launch control headquarters of the Strategic Air Command in Omaha, Nebraska, was deliberately isolated from the Internet. The scenario portrayed in the popular movie *War Games*, in which a teenage computer whiz-kid taps into the USA's nuclear arsenal from his home computer, was never possible in real life. The DoD built its own global communications network, the World-Wide Military Communications System (WWMCs, pronounced 'Wimmix'), which shared little with the Internet and was not connected to it; indeed, WWMCs was notoriously unreliable and was eventually abandoned.

For a variety of reasons, the development of the Internet, even when it was funded by the DoD, scarcely attracted the attention of military planners or national security officials. In the 1960s and 1970s, ARPA was an agency nearly unto itself, run primarily by and for academic researchers who were distant from military culture. ARPA's character did begin to change in the 1980s, but in the early days of the Internet, the system was viewed almost universally as a research programme, not as a precursor to a communications network tied to national security. In fact, it was this research character that contributed to the ease with which the Internet was absorbed by the civilian sector, and then by commercial enterprises. The Internet was not burdened with security classifications, secret budgets or secret technical specifications. And, ironically, it was this very openness of the Internet's development that reduced its importance in the eyes of career military officers and high national security officials, who were conditioned to believe that anything significant in their fields must be classified and secret.

In short, while the Internet and the concept of 'national security' share common roots in history, they developed along separate and divergent paths. This makes it all the more interesting that these paths are now converging again, but in a way that makes the Internet problematic – and even threatening – to national security.

THE NEW INTERSECTION OF NATIONAL SECURITY AND THE INTERNET

In September 1997, the President's Commission on Critical Infrastructure Protection released a preliminary report calling for a vast increase in funding to protect eight key elements of the US infrastructure: electric power distribution, telecommunication, banking and finance, water, transportation, oil and gas storage and transportation, emergency services and government services.[5] 'These are the life support systems of the nation,' said the commission's chairman, retired Air Force General Robert T. Marsh: 'They're vital, not only for day-to-day discourse, they're vital to national security. They're vital to our economic competitiveness world wide, they're vital to our very way of life ... The Internet provides an access point into all these infrastructures.' Commission member John T. Davis, representing the National Security Agency, said the government should develop a secure 'Next Generation Internet' for official use. The commission recommended doubling the current federal R&D budget of $250 million for protecting these systems, with increases of $100 million each year after 1999 to $1 billion per year by 2004.

In February 1998, US Attorney General Janet Reno unveiled a $64 million plan to build a new 'command center' to fight 'cyber attacks' against US computer systems. This new 'command center' is called the National Infrastructure Protection Center, and is a Justice Department response to the report from the President's Commission on Critical Infrastructure Protection.[6] These are just some of the more recent and visible results of concern over 'cyberwar', 'infowar', 'cyberterrorism' and other related threats now perceived by law enforcement personnel and national security officials as new and important terrain, and these authorities commonly view the Internet as the 'highway' upon which these threats will be borne.

As everyone knows, the character of the Internet has transformed dramatically in recent times. What began as a communications network for scientists, academics, engineers and specialists is now a vast global communications medium that rivals the public telephone network, television broadcasting, and even radio. The Internet-using population worldwide is now over 60 million people, and it has been predicted that by the year 2002 there could be more than 700 million people using the network. Senior executives in large telecommunications companies – an industry which is now the largest in the world – routinely report that data traffic will soon surpass voice traffic, and that packet-switched networks like the Internet may eventually supersede the circuit-switched telephone network worldwide. The Internet model of packet-switching, distributed communication and unmanned digital nodes appears to be the bedrock for nearly all future communications.

Of particular importance to those charged with national security is the fact that increasing levels of international commerce are conducted over the Internet, as are increasing levels of government service. International funds transfers, now surpassing over a trillion dollars a day, are carried by computer networks. Power grids, banks, government databases, large corporate enterprises, news networks, transportation facilities and many other essential components of civilized life are increasingly 'on the net', delivering services or conducting critical communications over the Internet.

It is feared that disruption of such services or communications could some day resemble or approach in severity an actual physical attack such as a military strike or a major terrorist incident. At present, the potential for a computer attack that would produce a major national calamity is controversial. Most computer attacks documented so far have been merely intrusions or annoyances. In many cases, vulnerability to computer attack is shrouded in secrecy or proprietary prudence. In other cases, vulnerability may be exaggerated to enhance the status and commercial value of computer security firms or to improve the negotiating position of government agencies that are seeking more funding or clout. What is important now, however, is that officials of the US Government and experts in the private sector are arguing persistently that the growth of the Internet, and its expanding capabilities, combined with the fact that it is increasingly embedded in 'critical national infrastructures', makes protection of computers on the Internet a matter of national security. In other words, regardless of the current threat, the future indicates growing vulnerability, and thus a growing urgency for protection and vigilance.

Jamie Gorelick, US Deputy Attorney General, told the host of the TV *Nightline* new talk show, Forrest Sawyer, in December of 1997: 'My own assessment, Forrest, is that we have a couple of years before there is a really serious threat. We have seen indications in criminal activity, in the plans of foreign nations, in the plans of terrorist groups that lead us to believe that we should be about the process of hardening our computers against attack.'[7]

As yet another irony, what may contribute to the threat of computer attack in the USA is the country's unrivalled military superiority. General Marsh said on the same *Nightline* programme: 'Nobody around the world today would attempt to defeat us on the battlefield. Instead, they will be seeking means to find vulnerabilities in our systems that they can exploit and do serious harm without having to confront us in the conventional armed way of the past.'[8]

If the Internet does prove to be a viable means for countries to attack one another, nations capable of such actions will find it far easier to afford a credible threat than would be the case with vast arsenals of missiles and tanks. A relatively modest investment in the skills of a handful of network trespassers and computer hackers would become a substitute for immense investments in weaponry. As such, the sources of credible threats could proliferate.

INTERNET CHALLENGES TO NATIONAL SECURITY

This 'new terrain' of computer warfare or cyberterrorism poses some serious and unfamiliar challenges to national security authorities.

The Challenges of the Internet

All forms of warfare in the past have involved a threat to geographically specific assets by equally geographically specific means, such as massed armies or ballistic missiles. One of the chief characteristics of computer attacks is their ambiguity in nearly every dimension: it is difficult to ascertain where the attack is coming from, who is behind it, what the motive is, whether it is the work of a determined enemy or merely a curious trespasser, and so on. Penetrations that come from trespassers inside the USA may not be benign or 'domestic'. Before the Gulf War, for example, there was a report of a US hacker breaking into Pentagon computers and then offering to sell the information to Saddam Hussein (who didn't buy it because he didn't believe it was genuine).[9]

It is not even clear what the term 'cyberwar' describes. If it means an organized and co-ordinated attack on computer systems by another national government, that may be too high a threshold; it is unlikely we will see an unequivocal example of this soon, except perhaps by the USA attacking an enemy's computers. 'Cyberterrorism' may be more likely, but as in the distinction between war and terrorism by other means, this prospect might call for solutions other than protection from 'cyberwar'. Roger Molander, an expert now employed by the RAND Corporation, says that if a computer attack were to occur in the midst of some other crisis of national security, the very ambiguity of the attack might complicate decision-making tremendously. This is a murky world for national security officials.

The USA has historically avoided major military attacks because of its relative isolation from belligerents, a kind of 'continental defense'. The country's military strategy has generally been to keep conflict as far from the US mainland as possible. But the Internet poses new problems. Its global character, and the way it works, allows easy access to almost any networked computer inside the USA, including those running critical systems, from nearly anywhere else in the world. For a determined adversary, there are now millions of entry points to the US heartland, and they require no logistical effort, in contrast to the obstacles facing adversaries in the past.

Because of the fact that computer attacks can come from both inside and outside the USA, and the fact that the origins of such attacks are difficult to identify promptly, jurisdictional controversies and overlaps among law enforcement and national security agencies are already rampant. The USA has had a long tradition, for fifty years at least, of separating the jurisdictions of agencies responsible for domestic threats from those responsible for foreign threats. If the Internet is factored into their responsibilities, these jurisdictional boundaries are rendered exceedingly vague and arbitrary, leading to confusion and conflicting interests.

Finally, the biggest issue of all: for the most part, in the past, the US military and its national security allies, such as intelligence agencies, have been charged with protecting military assets first, and using these as offensive weapons or deterrents against enemies. In a 'cyberwar' scenario, however, conventional military assets will be useless, and there may be no appropriate offensive or retaliatory weapons available. The military and law enforcement and national security agencies are increasingly faced with protecting private assets, such as corporate computer systems, or other information systems far outside the jurisdiction of the federal government. Given the nature of US democracy, the federal government's powers to force private companies or governmental entities to set up protection schemes are limited, and as demonstrated by the ongoing debate over encryption restrictions, the government may have interests quite different from those of private companies, especially those that compete in the global marketplace. Indeed, given the evolving nature of global enterprise, it is commonly unclear where a US company stops and a foreign counterpart or partner begins. The Internet does tend to erase national borders, as does global commerce. The US defence establishment has traditionally been able to circumscribe what constitutes a 'national asset', but this is becoming more and more difficult.

For all these reasons, many of which have emerged only in the past five years or so, the Internet is a new factor in national security assessment.

Given the significant influence of national security agencies in setting national political agendas and shaping technological trends, this new friction between the Internet and national security is likely to affect the way the Internet develops for the foreseeable future. At stake is whether the Internet can retain its democratic, global and egalitarian features, or whether it will be absorbed into older patterns of national competition for power and status.

How big a threat to national security is the Internet?

Unfortunately, the answer to this question is not clear. While advancing technology has made assessing all threats to national security increasingly difficult, for a variety of reasons, assessments of the threat of 'cyberwar' or 'cyberterrorism' via the Internet may be the most difficult of all.

In the first place, the Internet is constantly changing. Indeed, it may be the most rapidly evolving entity in human history. It is difficult, if not impossible, to freeze a 'moment' on the Internet to make an assessment that would be valid for more than a few weeks at most. This is very different to assessing other kinds of vulnerabilities or threats, which change or accumulate much more slowly. During the Cold War, US intelligence sources had a reasonable idea of the capabilities of the Soviet Union, at least in terms of the raw numbers of its military assets. It is difficult to imagine how the same sources could 'count' the threat of Russian hackers, for example, some of whom have penetrated deep into the computers of US-based banks, such as CitiCorp in New York. The Internet has also extended and deepened its reach so much over the past few years that it is almost certainly impossible for anyone, or any group of people, to know everything it touches at any given time. Not only is the system vast, involving tens of millions of computers, but it is characterized by rapid change, contingency, complexity, innovation and constant 'churning', or the birth and death of new features almost overnight. In short, this is a risk assessment team's worst nightmare.

Even if one were able to narrow one's focus to 'critical' systems connected to the Internet, there are no public or even readily available data on how vulnerable such systems might be. Defence computers are buried under layers of secrecy and classification, and private companies are not likely to volunteer such information. We typically only hear about computer vulnerabilities after a break-in, and even then we learn little about the incident, and sometimes the descriptions of break-ins are not even accurate. A New Jersey state trooper once told the press that a teenage hacker he had arrested was altering the orbital paths of US defence satellites, which was not only untrue, but absurd. People who reveal computer break-ins often have ulterior motives for such revelations. Responding to a recent rash of reported break-ins into Pentagon computers, Peter Neumann, one of the world's leading computer security experts, told *HotWired News*: 'Perhaps this is a con game ... You put out a system with miserable protection and hope that someone breaks it ... Then you can ask for millions of dollars more to perform further palliative protections, rather than getting to the core of the problem – significantly ratcheting up the security of the infrastructure.'[10]

When officials such as General Robert Marsh tell the press that the Internet provides access to many, if not all, of the critical infrastructure systems of the USA, it is difficult to assess this claim, except to suspect that he is right. The nature of the problem means it is unlikely to be detailed in government reports on the levels and sources of current risks. Accumulating evidence based on anecdotal reports is likely to be the only information available.

Because of the paucity of hard data, and the difficulty of assessing what computer systems are vulnerable because of being connected to the Internet, it is difficult to assess whether there is in fact an Internet-related threat that compares to other kinds of threats. Kevin Poulsen, who appeared on the *Nightline* TV programme mentioned earlier, and who was described on that programme as a 'former computer hacker', said: 'I've heard so much talk about the coming info war. I'll be more worried when somebody can actually show me a single case of a hacker doing something that was malicious. So far they haven't.' He went on: 'The most heinous, co-ordinated, planned-out, conspiratorial, hacker attack imaginable wouldn't come close to a single bombing of a building. Nobody's ever going to die from anything that happens electronically. The government has held me up as an example of a hacker that had reached the very top. If I wasn't anywhere near having that kind of capability, then what reason is there to think that anybody is?'[11] Most examples of computer break-ins have been annoyances and cause for alarm, not serious threats to critical systems. Gene Spafford, another computer security expert, likened hacker break-ins to: 'being pecked to death by ducks. No one of these instances is really serious ... But if you've got 10,000 people doing that, it's a huge problem.'[12]

There have been some worrying computer attacks, such as the attempt by German hackers to secure classified information and sell it to the Soviets, chronicled in Clifford Stoll's book *The Cuckoo's Egg*.[13] There is some evidence that there was an attempt to disable Croatian computers during its war with Serbia, and Croatian security experts suspected that Serbian programmers were responsible for the attacks, although there was no definitive evidence.[14] The 1989 Morris 'worm' that brought down thousands of Internet computers and the 1998 virus that affected *Windows* computers on the Internet highlighted the vulnerability of the system as a whole. The Pentagon has admitted that its computers have been penetrated hundreds of thousands of times. Federal officials have hinted in press briefings that they have classified information about far more serious hacking attempts, successes or penetration capabilities in other countries. And of course, the Pentagon is busy building an offensive 'infowar' capability of its own.

But the overall problem facing national security authorities is that this threat of Internet-based terrorism or attack, however grave it might be, is to date not at all tangible to the average citizen, nor is it likely to become more so in the near future unless a catastrophe occurs for 'demonstration' purposes. Their current strategy is to request vast sums of money to prevent something occurring, not unlike the 'millennium bug', which the public also barely understands, if at all.

This is again far different from the world of the recent past, in which the threat of a Hiroshima-like nuclear explosion taking place in the USA was lodged quite vividly in the minds of most citizens. It is considerably more difficult to persuade the public that there is a large potential for threat to the nation via the Internet, when the entire country is involved in a massive campaign to get everyone on-line, especially schoolchildren, and when there is no obvious way to identify the full nature of the threat. Once again, this is new terrain for national security advisers.

As everyone knows, 'national security' is largely a game of perceptions, a combination of both real and imagined threats and assets. Even during the Cold War, there was controversy over how big a threat the Soviet Union actually posed to the USA. This controversy continues even today, over ten years after the end of that conflict, so it is not surprising that there is a debate over whether there is a national security threat posed by the Internet, or whether this is paranoia, or, more cynically, whether this is related primarily to institutions hoping to increase their budgets and their longevity. This controversy is fuelled by the sparse information available about the true level of the risk, especially with respect to 'critical' systems. Because we can expect this dearth of information to continue, the controversy about the nature of the threat will no doubt extend far into the future as well. In the digital era, the very nature of the technology paradoxically makes perceptions more important, because facts are harder to come by.

What do we know? When computer security experts are asked whether Internet-networked computers are secure, their answers are almost always along the lines of 'not enough', or 'not yet'. One might discount such answers as being motivated by self-interest and still conclude that more needs to be done about computer security. The explanation given by security experts about why we don't do more is that the public has not yet demanded greater security for computers, and the lack of significant public demand means that companies are not providing it. It is also expensive and sometimes troublesome to secure a computer and to keep it secure, imposing discipline on users and system administrators who would rather not be disciplined. It is common to hear of people learning about computer security the hard way, in a 'trial by fire', absorbing a lesson after something nasty has occurred. Obviously, if there is a real threat to national security via the Internet, such lessons are not an adequate substitute for prudent policy-making. It is the job of national security officials to prevent catastrophes, not to say 'I told you so.'

The vexing issue is how we might feel safer without seriously compromising the best features of the Internet, trampling on democracy or turning into a surveillance society. These are not new concerns: they were not introduced into public debate by the appearance of the Internet. But they have been made rather dramatically more complicated by the first truly significant supranational sphere of discourse and politics. They are further complicated by the dual role of national security agencies: to protect national assets, and to penetrate the defences of enemies. It is this dual role, embedded in the traditions and histories of national security agencies, which is at the heart of the intense debates about a possible solution to computer-based threats: widespread digital encryption.

THE INTERNET AND NATIONAL SECURITY AGENCIES

By now, nearly every federal agency within the US Government has some department or division responsible for computer security. But the preeminent agencies in this field are still those charged with national security, such as the National Security Agency (NSA), the Central Intelligence Agency (CIA), the Federal Bureau of Investigation (FBI), and to a lesser extent, the Justice Department and the Secret Service within the Treasury Department. It is important to acknowledge that it is the character of these agencies, their histories and their own responsibilities that give the subject of computer security in the USA a particular kind of atmosphere, largely that of the military and national security community itself. Thus, the 'command and control' model of computer security has tended to dominate the US Government's approach.

In 1987, the Computer Security Act of the US Congress apportioned responsibilities for computer security to the National Institute of Standards and Technology (NIST) of the US Department of Commerce for non-classified computer systems, and to the NSA for classified systems. This law was the result of a certain level of alarm, on the part of Congress and civil libertarians, during the Reagan administration because of a pair of White House national security directives that pointed towards NSA control over all computers in the USA. The Computer Security Act was an attempt to mark a boundary for civilian control of unclassified information systems.

However, since the Computer Security Act, the NSA has worked diligently to regain and secure its supremacy over computer security policy. A 1989 'Memorandum of Understanding' between the NSA and NIST shifted power back to the NSA, and in 1994 President Clinton issued Presidential Decision Directive 29, setting up the Security Policy Board, which has recommended that all computer security functions for the government be merged under NSA control.[15]

At the same time as the NSA has attempted to impose its own standards on computer security in the USA, the FBI has tended to extend its responsibilities beyond domestic law enforcement to international crime, counter-terrorism and counter-intelligence. While officials of the NSA are largely unknown to the public, FBI Director Louis Freeh is a common face in the news, often called upon to testify and make the government's case for control over encryption and computer security in the name of national security. The 1993 bombing of the World Trade Center in New York City, which was apparently connected to a foreign conspiracy in the Middle East, strengthened the FBI's role in monitoring international terrorism immeasurably. Freeh also points to international drug cartels, new foreign sources of organized crime, the international terrorist activities of so-called 'rogue states', the frightening potential for the uncontrolled proliferation of small nuclear weapons and other threats to make the case that the FBI now has a host of new targets.

These two phenomena of recent years have tended to blur the line between domestic security and national security, which has produced crises of constitutional protections in the past. But Congress has been persuaded by the federal law enforcement and national agencies. The Intelligence Authorization Act of 1997 states: 'elements of the Intelligence Community may, upon the request of a United States law enforcement agency collect information outside the United States about individuals who are not United States persons. Such elements may collect such information notwithstanding that the law enforcement agency intends to use the information collected for purposes of a law enforcement investigation or counterintelligence investigation.'[16]

Whitfield Diffie, the co-inventor of Public Key Encryption, and his co-author, Susan Landau, in their 1998 book *Privacy on the Line*, comment: 'This wording carefully steers clear of permitting the intelligence community to spy on Americans directly, but opens the way for unprecedented collaboration between the intelligence and law enforcement communities.'[17] The boundary-free character of the Internet will very likely intensify the merging of conventional law enforcement activities inside the USA and the activities of national security agencies. The front line in the debate about this trend is encryption policy. In September 1997, the National Security Committee of the US House of Representatives voted to strengthen controls on the export of digital encryption, reversing a trend toward relaxing such controls, one of the chief goals of the high-tech industry. Committee members cited the warnings they received in 'classified briefings' as the main reason for their decision. Congressman Mike Oxley, a Republican from Ohio, told the *New York Times*: 'I would find it difficult to believe that a member who heard the briefing could walk away not committed to address-ing security issues. Frankly, I wish everyone interested in this issue could have heard for themselves the alarming briefing that members of our com-mittee heard.'[18]

Constraints of space prevent a comprehensive or even adequate review of the encryption debate here. In summary, the US Government's position is that law enforcement and national security authorities need to retain the ability to intercept and interpret communications, including digital data, in order to fulfil their responsibilities to protect the USA from crime and foreign threats. As such, the federal government has proposed a series of arrangements that have all included the concept of 'escrowed keys', mean-ing that authorized officials will be able to acquire an encryption key to unlock scrambled data. Opponents of this scheme, which is the traditional approach in military cryptography, argue that Public Key Encryption, with-out escrowed keys, is the only safeguard for privacy and authenticated electronic commerce. These opponents also argue that criminals and foreign adversaries will have no incentive to use encryption with escrowed keys – the 'escrowed key' approach will provide keys to the communications of people who obey the law, while others will have easy access to unbreakable encryption algorithms. The ready availability of Public Key Encryption, argue government critics, means that 'the horse is out of the barn' already. Moreover, they point out, the task of security in the computer age is for each person, or each computer administrator, to be responsible for computer security, because the task is too immense and complex for bureaucratic oversight. The argument has been framed as one in which the protective schemes are characterized as a choice between armies or locks, and each of these has its attendant interest group.

While Congress has been largely sympathetic to and supportive of the federal law enforcement and national security agencies, another front has opened up in the US courts. In early 1997, in a case heard before the US District Court for the Northern District of California, *Bernstein* vs. *United States Department of State*, US District Court Judge Marilyn Hall Patel ruled that national security considerations cannot be used to censor cryptographic schemes on the Internet. Daniel Bernstein, a graduate student, created an encryption algorithm called 'Snuffle'. After four years of correspondence with the US State Department, Bernstein learned that his source code and all other material about 'Snuffle' except a research paper were in violation of export control laws and could not be posted to the Internet. Bernstein sued the State Department, claiming that this ban violated his First Amendment rights. Judge Patel agreed with Bernstein, and wrote a stinging rebuke of the government's position. Patel ruled that computer source code is protected as free speech by the First Amendment, and that the government's attempt to ban such speech amounted to 'prior restraint', which is unconstitutional. Judge Patel specifically prevented the US Government using claims of a breach of national security to impose prior restraint on the distribution of encryption source code.[19]

The US Government lost its appeal of the Bernstein decision in the US Circuit of Appeals in San Francisco, but is expected to carry its appeal to the US Supreme Court. If the Supreme Court upholds Judge Patel's ruling, this may close the encryption debate for the foreseeable future; after such a ruling, all encryption source code could be posted to the Internet without intervention by national security or law enforcement agencies. This would effectively kill the means by which such agencies now control the distribution of non-escrowed encryption algorithms. On the other hand, if the Supreme Court overturns the Bernstein decision, this will reinforce the role of national security agencies in shaping the future of the Internet. Because of this, the Bernstein case is being watched very closely by both sides of the encryption debate.

The dispute over encryption has put into stark relief the dual nature of the national security mission: the responsibility of such agencies to protect national technological assets while retaining the ability to intercept and interpret digital communications. In the era of the Internet, these two responsibilities are in conflict, thus posing significant dilemmas for national security officials. On the one hand, these agencies are urging businesses, other government agencies and individuals to protect their computer data from attack. On the other hand, they seek to control the way these people protect such data, in order to ensure that law enforcement and national security agencies have access to this data. Not surprisingly, both businesses and individuals are hesitant to implement encryption schemes that require them to turn over keys to people they don't know or whose motives they don't fully comprehend. Because of this hesitancy, the first mission of the security agencies – increasing protection for US computer systems – is stymied.

General Marsh, the Chairman of the President's Commission on Critical Infrastructure Protection, has told the press that he hopes the dispute about encryption policy is resolved soon, because he believes, with justification, that this ongoing dispute is obstructing the implementation of greater security for computer systems. Members of the commission have hinted at their support for a loosening of encryption controls, but have also leaked the information to the *New York Times* that 'they are under fairly strict orders to fall in line with the FBI's push for key recovery'.[20]

Business executives are understandably reluctant to invest in security systems that may be superseded by other technologies or blocked by court rulings. The US Government's rapid transition from one policy recommendation to another – such as from DES (Data Encryption Standard) to the 'Clipper chip' to 'key escrow' to 'key recovery' – has not helped foster confidence in the business community. And of course, some influential business organizations, such as the Business Software Alliance,[21] are allied with civil libertarians and other opponents of US Government policy in the case of encryption standards.

Despite the hopes of General Marsh and others, at present the encryption debate in the USA is so polarized that compromise solutions are not visible, nor likely to emerge soon. This polarization has been exacerbated by some ideological and political trends in the USA; Republican conservatives, who now control the US Congress, are more sympathetic to law enforcement and national security arguments than was true of Congress just a few years ago. For example, for many years the US House Subcommittee on Civil and Constitutional Rights was chaired by Don Edwards, a Democrat and a former FBI agent who was highly critical of any law enforcement encroachment on civil liberties. But Congressman Edwards has retired, and a conservative Republican took his place as committee chairman.

Another new phenomenon is the emergence of a new breed of 'cyberlibertarians' – typically young, talented technologists who reject most of the assumptions of the national security establishment. Some of the more radical of these ideologues have argued that the Internet is the beginning of the end of the nation state, not just the end of the national security state. This perspective is not an isolated intellectual discourse: the 'cypherpunk' movement, for example, has complex and extensive ties to outlaw hackers. Most of these hackers are young men, some of whom have adopted the intellectual framework of 'cyberlibertarianism' as an ideological justification for criminal penetrations of government computers – casting themselves as 'Thomas Paines' of the digital revolution. Such 'counterculture' attitudes are widely shared by educated young people all over the world – perhaps a natural attitude of young people rebelling against authority, but in this case these people are among the most technically adept in the world, and a number of them are even wealthy because of this skill. Hence, once again, national security officials are confronting a new and alien environment, one dramatically different from eras of the past, when business leaders and skilled technologists were typically undisturbed by the alleged imperatives of national security. Now, when confronting young leaders of the digital revolution, national security authorities are in hostile territory. The end of the Cold War has given new impetus to calls for the dismantling of national security institutions, and the Internet, with its idealistic potential for global communications between planetary citizens, has come along at just the right time to fuel such ideas. Widespread use of phrases such as 'the digital revolution' and 'Third Wave civilization' (lifted from the work of the Tofflers, and adopted by the ex-Speaker of the US House of Representatives, Newt Gingrich) reinforce the popular notion that the 'information age' entails an overturning of old regimes, including perhaps an end to the centuries-old competition between nation states.

For these reasons, in addition to the prosaic clash of interests between the government and pragmatic corporations that are part of a global marketplace, the encryption debate is the leading edge of a much larger philosophical debate about the role of the state in the information age. It is unfair, of course, to characterize national security authorities as 'dinosaurs' due for extinction, the way they are characterized by some of the 'cyberlibertarians'. Even a small dose of the daily news is enough to convince most people that there are real threats that continue to justify the need for some national security protections. On the other hand, there is no compelling reason to assume that the missions and structures and traditions and size of national security institutions created during the decades of the Cold War need to persist into eternity. Many critics of national security agencies argue reasonably and persuasively that the Cold War should be regarded as an anomaly in US history, a struggle that imposed sacrifices in democratic values that need not be sustained or repeated in the absence of a threat equivalent to the Soviet Union. These critics have quite rightly put on the table for debate that question of whether the 'national security state' is an essential or necessary political form for nations in the twenty-first century, particularly in an era in which the Internet is challenging many assumptions and norms inherited from the pre-Internet period, the Second World War and the decades of the Cold War.

It may seem grandiose to suggest that complex technical debates such as those surrounding digital encryption or the prospect of 'cyberwar' are the most important political debates of our time, but this is the case. This is not all that surprising when one considers the catalyst for such debates: the Internet itself, one of the most remarkable, promising and at the same time vexing creations of human enterprise in the history of the world.

RECOMMENDATIONS

National security authorities, like everyone else, are confronted with a world far different than the familiar one of just a few years ago: the combination of the end of the Cold War, the new intensity in global commerce and competition, and the information revolution has served to up-end almost all previous cognitive models about how the world works. National security bureaucracies are notoriously resistant to change, but they also have many arguments on their side, as new threats have appeared alongside new ways of communicating and doing business.

National security experts are facing several frustrating dilemmas. First is the need to secure US computer systems while retaining some ability to intercept and interpret digital communications. As many people have pointed out, this dual mission may not only be irreconcilable, but the effort may produce some absurdities and unacceptable impositions on people using computers or developing information technologies, particularly software. Peter Wayner, a reporter for the *New York Times* who wrote about the Security and Freedom Through Encryption (SAFE) Bill considered in the US Congress in 1997, wrote: 'The bill would force developers of new software to seek approval for their products from the United States government even if the products did not explicitly include encryption features. Such approval would be the only way to escape prosecution, [a Congressional staff member] said. While admitting that this language would add a six- to nine-month delay in releasing new products, the staff member asserted that the computer industry would simply have to build this time into product development cycles.'[22] Given the life spans of computer software today, and the prospect of international competition unburdened by such delays, the imposition of a nine-month delay in releasing software products appears fatally ill-advised. The idea that computer software companies might need to have their products reviewed by the government, like prescription drugs, is also alarming and bizarre: the task would most likely prove impossible, not to mention absurd.

Thus the implications of several initiatives by the national security community are so onerous and so out of touch with the imperatives of the digital economy that their chances of becoming law in the USA appear to be slim. Lawmakers are loath to alienate national security authorities, but in this case they may have no choice – the economic health of the USA could be seriously damaged by several of the proposals now on the table.

Moreover, it appears inevitable that uncontrolled Public Key Encryption algorithms will proliferate, despite the resistance of the US Government. 'Key escrow' systems such as those advocated by US Government officials are too vulnerable to compromise, and once keys are released into circulation, all encrypted data are compromised. Public Key Encryption schemes are already available in a wide variety of products and on the Internet, and there appears to be little the government can do about these programs. If the Bernstein decision is upheld by the Supreme Court (and this court has been consistently vigilant about challenges to the First Amendment), it will be illegal – unconstitutional – to block the distribution of source code, rendering all the efforts at government control moot. It is also significant that foreign governments do not share the US Government's position on encryption, which creates a vast 'safe harbour' for alternative encryption schemes that, because of the way the Internet works, would be merely 'a click away'.

The arguments of proponents of Public Key Encryption, such as Diffie and Landau, are generally persuasive. They maintain that communications interception is a 'low-value' activity of law enforcement and national security agencies, and is far outweighed by the value of more secure computer systems throughout society. They point out that surveillance of foreign communications is dependent on agreements on 'escrow' keys between foreign surveillance targets and the US government – a rather improbable scenario. Diffie and Landau ask what the consequences would be if policy-makers were to regulate encryption.

If cryptography comes to present such a problem that there is a popular consensus for regulating it, regulation will be just as possible in a decade's time as it is today. The laws could be changed, strong cryptography would no longer be incorporated in new products, and the ready availability that government claims to fear would decline again quickly. If, on the other hand, the precedent of building government surveillance capabilities into our security equipment were established, the very survival of democracy would be at risk. Diffie and Landau say: 'government efforts to keep honest citizens from using cryptography to protect privacy continue. Such efforts are unlikely to achieve what governments claim to want, but very likely to cause serious damage to both business and democracy in the process.'[23]

Widespread use of Public Key Encryption appears to be the only viable and cost-effective means of truly securing computer systems essential to the functioning of modern society. The task then becomes one of adjusting the roles and activities and mind sets of national security officials to this fact. This needs to be a process of collaboration, as opposed to the current process that is characterized by polarization, hostility, suspicion and even attitudes that suggest that each side wishes the other side would die off and fade into obscurity. There is no precedent for collaborative work between citizens and national security agencies in the USA. There is a long history within such agencies that can be summed up in the phrase: 'If only you knew what we know, you'd agree with us.' But then, of course, the knowledge referred to is out of bounds, unavailable, secret, incapable of being assessed except by those deemed trustworthy enough to possess it, which typically means people who already agree with the assumptions of the intelligence and national security communities. This has to change somehow, and the end of the Cold War opens up a historic new opportunity.

The US Congress should take the lead. Members of Congress should understand the stakes. In the case of the intersection of the Internet and national security, the perspective of national security agencies is only one side of the coin, maybe even vastly overbalanced on the other side by the potential damage to business and democracy. Unfortunately, Congress has a history of being cowed by national security briefings. Congress needs another leader like Don Edwards, who was not intimidated by officials of the FBI or the CIA. Whether this occurs remains to be seen. High-technology executives need to understand the need for such a leader, even if such a person doesn't agree with other features of the high-tech sector's public policy agenda.

The public is not likely to be a major player in this debate. The subjects of national security and technology have always been reserved for elites, and while this situation may be regrettable for democracy, it should not be expected to change soon. Consequently, there needs to be intense work on the part of the business community to persuade the White House and Congress that the world is now a different place, that the changes recommended by government authorities are dangerous and unworkable, and that business stands ready to co-operate with national security officials in finding new solutions. To a certain extent this is already going on, but the dialogue with national security officials could be improved significantly if their Cold War rhetoric was attenuated, or even abandoned.

President Clinton could be the leader this issue needs, but unfortunately he doesn't appear to be up to the task. He is a president more than usually shaped by the demands of law enforcement and national security authorities – the Clinton administration has been one of the worst in recent memory for civil liberties in the USA. The president is also famously averse to friction and confrontation, despite the omnipresence of these qualities during his period in office. He is probably not going to do anything to alienate either the national security community, upon whose approval much of his stature as a 'law and order' president depends, or the high-tech community, many of whom helped get him elected. The president's dithering on the issue of encryption may thus set the terms of the debate – a kind of policy fibrillation – until someone else holds his office or some other leader finds the means for a real breakthrough.

In the meantime, proponents of both sides will continue to find opportunities for strategic advantage. Professional societies such as the Internet

Society, the Association for Computing Machinery and the Institute of Electrical and Electronics Engineers should increase their efforts to find a workable solution. They might also consider making efforts to educate the public about what is at stake, perhaps by sponsoring television programmes or a national campaign of public debate and community meetings. The high-tech industry has every reason to help fund such public outreach efforts.

Finally, professional societies need to do more to educate their members that the most fundamental interests of the computing profession are not primarily about technical issues, but are tied up in public policy controversies. The computer industry has demonstrated time and time again that technical obstacles can be overcome – often with astonishing, disorienting rapidity. What delays progress in the information age are social and political controversies that have fewer, if any, black-and-white answers. People in the computer industry need to become far more sophisticated about policy issues, political participation and how technology affects basic values in society. For too long, public policy has been considered a field separate from technology, and of only marginal importance. This may be changing, but if so, it is changing too slowly to keep pace with the issues confronting us now. Other technical and scientific fields, particularly physics, do a much better job of integrating public policy work into their professional activities. There are lessons in the experiences of physicists that might help computer professionals, especially because of the tight coupling of the work of physicists with the field of national security.

CONCLUSION

It was inevitable that the global civilian extension of the Internet – a natural and predictable development of computer networking – would clash with traditional principles of national security. However, admitting this with the benefit of hindsight does not relieve the pressures of this clash that exist today, many of which are so vexing as to seem nearly insoluble. Two immense forces of great momentum are at odds: technological progress, which takes a million different forms, emerging from countless points around the world; and national security, the gravest and most fundamental public responsibility of the world's richest and most powerful nation. How the frictions between these forces will be resolved is not yet clear. Neither is likely to go away, or even fade in strength.

What seems to be required is a new concept of 'national security' that can accommodate the Internet to both civilian and national security interests. This does not have to be a radically libertarian Utopia, in which the nation state itself withers and dies, as seems to be the hope of some young cyber-activists. Nor does it need to be an accommodation in which national security and police surveillance and enforcement are the rulers of the Internet. Any new accommodation would probably need to take uncontrolled public key, so-called 'strong' encryption for granted, as this seems to be inevitable. National security authorities were once faced with another technological revolution of comparable significance – intercontinental ballistic missiles with nuclear warheads – and security policy managed to adjust, for better or worse, to this new technology. The same kind of adjustment will now be required. The 'national security state' that was a product of the Cold War may no longer be recognizable in ten or twenty years, but neither will any other institutions of modern society, because of the changes tied to the Internet. Because of this, national security officials need to start thinking in fresh ways. Right now, they are on the wrong side of history, as noble as their aims may be.

REFERENCES

1   Katie Hafner and Matthew Lyon, *Where Wizards Stay Up Late: The Origins of the Internet* (New York: Simon and Schuster, 1996).
2   Ibid., p.54.
3   Ibid., p.64.
4   Ibid., p.77.
5   See, for example, the special issue 'Information Warfare', *Aviation Week and Space Technology*, Vol. 148, no. 3 (19 January 1998), pp.52–60.
6   James Glave, 'Critics bash Reno's cyberwar plan', *HotWired News*, 27 February 1998. Available at: <http://www.wired.com/news/news/technology/story/10605.html>.
7   *Nightline* (ABC News Talk talk show), 9 December 1997. Transcript available at: <http://www.infowar.com/CLASS_3/class3_011298a.html-ssi>.
8   Ibid.
9   Ibid.
10 Glave, 'Critics bash Reno's cyberwar plan'.
11 *Nightline* transcript.
12 Glave, 'Critics bash Reno's cyberwar plan'.
13 Clifford Stoll, *The Cuckoo's Egg: Tracking a Spy Through the Maze of Computer Espionage* (New York: Doubleday, 1989).
14 Personal interview with the Croatian Deputy Minister of Science and Technology, Predrag Pale, 16 March 1998, Zagreb, Croatia.
15 EPIC (Electronic Privacy Information Center), 'The Computer Security Act of 1987', available at: <http://www.epic.org/crypto/csa/>.
16 Ibid.
17 Whitfield Diffie and Susan Landau, *Privacy on the Line: The Politics of Wiretapping and Encryption* (Cambridge, MA: MIT 1998), p.123.
18 Peter Wayner, 'Computer privacy: Your shield? Or a threat to national security?', *New York Times*, 24 September 1997.
19 Guylyn R. Cummins, 'National security alone can't be used to censor cryptographic

speech on the Internet', *Daily Transcript*, 12 March 1997. Available at: <http://www.sddt.com/reports/97reports/iWorld97_03_12/DN97_03_12_ti.html>.
20   Wayner, 'Computer privacy'.
21   <http://www.bsa.org>.
22   Wayner, 'Computer privacy'.
23   Diffie and Landau, *Privacy on the Line*, p.245.

3 Defence Diversification in the United Kingdom

Ian S. Goudie

INTRODUCTION

Since the end of the Cold War, most Western governments have announced reductions in defence requirements. By 1997–98, defence spending in the UK was planned to fall in real terms by some 20 per cent since the end of the Cold War.[1] As the size of the defence establishment is decreasing, there are both potential costs and benefits to the civilian sector of the economy. These include the transfer of property, technology and expertise from one sector to another. This chapter will examine the problems of this transfer for the UK, as one example of what is happening globally in developed countries.

Among the potential benefits of the reduced defence requirements is the re-sale of defence estates.[2] The UK Ministry of Defence (MoD) owns around 227,000 hectares and leases a further 16,000 hectares. This is 1 per cent of the total area of the UK, making it the second largest estate in single ownership in the UK after the Forestry Commission. In addition to the Defence Estate proper, the MoD holds rights of various sorts and for various terms over a further 270,000 hectares of land, principally for Army training, including 55,000 hectares in Northumberland and 45,000 hectares in Tayside. This is a further 1 per cent of the land area in the UK.

Many localities around the country include MoD land or facilities. This varies from 41,388 hectares in Wiltshire to 1 hectare in both the Borders Region of Scotland and in West Glamorgan. While the size of the estate has remained almost static for the last two decades, it is now in the throes of major change. As the MoD budget is shrinking, much of this land is being disposed of at 'market rates' to the civilian sector.

Unemployment is a major cost incurred by the civilian sector due to the decrease in demand for military industrial production. The number of UK service personnel was planned to fall from 306,000 in 1990–91 to around

217,000 in 1996–97 – a reduction of 29 per cent. UK-based MoD civilian staff numbers were projected to fall from 129,000 in 1990–91 to 111,000 by 1996–97 – a 14 per cent reduction. The 'First Line First' review of defence administration and support proposed additional cuts of 7.5 per cent in the civilian staff and 5 per cent in the armed forces by 1997–98. A wide range of base and other military establishment closures were also proposed.

The MoD service and UK-based civilian staff in 1998 was about 400,000, compared to 487,000 in 1990. The MoD has estimated that in 1992–93, some 425,000 jobs in the UK depended on MoD expenditure on equipment and other goods and services, and on defence exports; 237,000, or 25 per cent of UK directly defence-dependent jobs have gone since 1990.

In the UK, a consensus has been developing over recent years which has called for greater networking between firms, and between key agencies such as the MoD, the Department of Trade and Industry (DTI), the Office of Science and Technology (OST), industry, trade associations, academia, local authorities and trade unions, to help attain a higher state of industrial competitiveness and stimulate innovation.[3] The Technology Foresight Defence and Aerospace Panel recommended a co-ordinated approach to the management of change within the defence and aerospace sectors.[4] A Defence Diversification Agency (DDA) operating at UK and national/regional levels would provide the required body to link the various agencies and initiatives while promoting defence diversification.

This chapter assesses the situation facing the defence industry in the UK. It provides a detailed proposal for the establishment of a specific agency responsible for ensuring that defence-dependent companies and communities have every opportunity to enter into non-military activities and to transfer some of their efforts from the military to the civilian sector. As the pace of defence restructuring on the European level increases, this proposal for a Defence Diversification Agency could provide the basis of a model to ensure that defence workers and their communities across Europe have the opportunity to benefit from a 'peace dividend', rather than paying the economic and industrial costs of a 'peace penalty'.

THE DEFENCE DIVERSIFICATION AGENCY

The UK Labour Party was elected to government in May 1998 on a manifesto which included the establishment of a Defence Diversification Agency. The government believed the defence industry's expertise could be extended to civilian use through a DDA, but were open to suggestions about how this could be achieved. However, the Arms Conversion Project and the Scottish Trade Union Congress had been working on the issue for some time, and had presented a draft working paper on such an agency to the Labour Party's Shadow Secretary of State for Defence, Dr David Clarke, in December 1997, who welcomed their efforts as 'the most comprehensive proposals on defence diversification'. The Secretary of State for Defence, George Robertson, speaking at the Arms Conversion Project Seminar held in Glasgow in early March 1998, supported this opinion, stating: 'I know that Ian Goudie's Arms Conversion Project already has a draft Working Paper on a DDA. This paper, which I have personally studied, will have a powerful input into the debate on our future plans for a Defence Diversification Agency which we hope to have in place by the year 2000.'[5]

A few days after George Robertson's speech, the government published its Green Paper entitled *Defence Diversification: Getting the Most Out of Defence Technology: Proposals for a Defence Diversification Agency*.[6] Following consultation, a White Paper was published on 5 November 1998 setting up the agency.[7]

The DDA is an important UK Government initiative which aims to gain greater benefits for small to medium-sized businesses from the nation's investment in research and development within DERA (Defence Evaluation and Research Agency). It encompasses a range of developments already under way at DERA sites, including dual-use technology centres, science parks and innovation centres. The DDA will have an operational budget of some £2 million a year, to be found from within the defence budget. It will provide a point of entry for those organizations that have little experience of working with DERA, allowing them to receive advice and gain information on the technology available, thereby facilitating access to DERA laboratories, stimulating transfer of the MoD's intellectual property rights and encouraging partnerships for co-development programmes.

The agency is one of the key elements of a series of complementary initiatives by DERA designed to create greater benefits from the nation's investment in research and development. These include establishing six dual-use technology centres and encouraging the development of science parks at a number of sites. DERA has already made significant contributions to the success of a number of non-defence companies in a wide range of areas from pharmaceuticals to racing car design.

The DDA will link its efforts with other existing government and European initiatives. Small and medium-sized enterprises already have access to a number of sources of business support, and the DDA is keen to avoid wasteful duplication or unnecessary confusion by establishing yet another independent business support agency. The DDA will build on the experience DERA has gained by working closely with the emerging regional development agencies, government offices, Business Links and other business support organizations to ensure that its activities complement those which are already well established.

In June 1999, the government announced the appointment of Professor Damien McDonnell as Director of the Defence Diversification Agency, supported by a number of Technology Diversification Managers based at DERA sites throughout the UK. The DDA is scheduled to be fully operational by January 2000. The agency will need to devote early attention to the most effective way to combine its efforts with other initiatives such as Foresight, the European Union's KONVER II programme (which will end at the end of 1999), and any relevant structural fund programmes after 1999. It will make a contribution to the competitiveness of a knowledge-driven economy in which creation, dissemination and exploitation of knowledge from defence research will be an important component. While the DDA will only deal directly with diversification as a means of technology transfer from DERA, rather than the diversification of companies and communities, these wider defence diversification issues have in part been recognized by the planned creation of a Defence Diversification Council which will include representatives of industry, local authorities, trade unions, the Department of Trade and Industry and the Scottish Parliament and Welsh Assembly.

THE CURRENT DEFENCE AND AEROSPACE SITUATION IN THE UNITED KINGDOM

The UK defence and aerospace sector

Defence and aerospace is a sector in which the UK is currently a world leader, with a long history of excellent science and product innovation. The UK is one of a very few countries with the capability to design, manufacture, integrate and market complete sea-, land- and air-based aerospace systems: fixed- and rotary-wing aircraft, aero-engines, air-traffic control systems, warships and submarines, air-to-air and air-to-surface missiles, low-level air defence, field guns and military land vehicles. There is also a strong and diverse equipment sector providing sub-systems and components worldwide. Many sub-systems have great technological sophistication and complexity, and high added value.

Defence and aerospace is also a sector in which science and technology are particularly important to competitiveness. Indeed, defence and aerospace has one of the highest research and development (R&D) intensities of all major UK industrial sectors – in 1992, the UK defence and aerospace industry R&D amounted to £1.63 billion, representing 25 per cent of the total R&D undertaken by UK business in all sectors. In 1998, aerospace R&D in the UK was at a level of 9.5 per cent of gross sales and defence R&D was 14 per cent of gross sales, compared with 2.2 per cent for industry overall. The electronics industry is the only other UK sector with a comparable R&D intensity at 12 per cent of gross sales. The defence and aerospace sector's strong dependence on science and technology makes it a good vehicle for UK wealth-creation, given that scientific research is a particular strength in that country.

Over 300,000 people are directly employed in the defence and aerospace sector (roughly 7 per cent of the UK manufacturing workforce), and at least another 300,000 jobs are generated indirectly by it. The sector accounts for 6 per cent of the UK manufacturing output, and over the past ten years has grown significantly faster than UK manufacturing as a whole.[8]

The UK's share of OECD (Organization for Economic Cooperation and Development) exports is higher for defence and aerospace (at 11.7 per cent) than it is for any other sector.[9] Defence and aerospace companies are among the UK's top exporters: in the UK table of exporters of 1993, British Aerospace occupied first position, Rolls-Royce occupied third position and the General Electric Corporation occupied fourth position.[10] Defence exports in 1993 totalled some £7 billion, giving the UK over 16 per cent of the market, second only to the USA as a defence exporter. Aerospace has the highest export ratio (60 per cent) of all the main UK industrial sectors, and has consistently achieved a positive balance of trade of between £2 billion and £3 billion annually. Only one other sector has a better trade balance in recent years, namely chemicals and pharmaceuticals. Significantly, this is another sector in which government-funded R&D plays a key role.

Change in UK defence and aerospace markets

The end of the Cold War has brought heightened threats of regional conflicts, proliferation of weapons of mass destruction and increased demand for peacekeeping and humanitarian actions. Whereas Western force structures have reduced, the defence market is growing in the newly developing nations, especially in the Asia-Pacific region.

However, the civilian aerospace sector has grown at an average rate of 6.2 per cent annually over the past twenty years, driven by a rapid growth in air travel. In most forecasts, this growth is predicted to continue, particularly because of the projected increase in air travel in the developing nations. Here, technology will contribute to both product performance improvements and product competitiveness. The UK has historically maintained over many years a roughly constant 10 per cent share of the civilian aerospace market. If the UK could achieve a modest expansion in its share of this growing market, the increased turnover would make a significant contribution to UK growth over the next twenty years.

Technology development in the UK defence and aerospace sector

Defence and civilian aerospace are global industries. They are high-value and relatively low-volume markets, requiring leading-edge technologies. The requirements for state-of-the-art technology provides a major stimulus for product and technology development from other sectors ('spin-in'), and the defence and aerospace sector often provides technology, products or processes which can be exploited by other sectors ('spin-out' or 'spin-off'). Increasingly, the defence sector will be looking for technologies with true dual-use potential, where the cost of technology development can be shared with partners requiring similar capabilities.

In defence and aerospace, more than in any other sector, the national government has a crucial impact on national competitiveness. Government is the predominant market force, being the main customer for the UK's defence industry and its principal source of R&D funding, and has been assigned statutory responsibility for the health of the UK civilian aerospace industry through the 1982 Civil Aviation Act. The success of the industry derives largely from the previous levels of investment by industry and government in research, development and procurement during the 1970s and 1980s.

A special feature of the defence and aerospace sector is that new product developments are characterized by long time scales and high costs, requiring large long-term investment. For example, the Eurofighter Aircraft (EFA) concept dates from 1983, but Eurofighter *Typhoon* will not enter service before the year 2000 and is likely to remain in service until 2030–40. Technology incorporated into new products may have a lifespan of 25–50 years. For example, the Eurofighter is a product whose technologies have their origin in research carried out in the 1960s and 1970s.

These very long time scales make the sector relatively unattractive for private-venture investment, so government assistance has been crucial. However, to date there has been relatively little civilian spin-off from this government assistance. Hence, the proposed Defence Diversification Agency should seek to ensure that the potential for civilian spin-off is maximized in future defence procurement contracts. In the past, for a defence company to invest speculatively to develop new non-defence-related equipment was regarded as unacceptably risky. However, defence companies have a responsibility to try to change the prevailing culture by persuading their shareholders that a larger proportion of profit must be reinvested in civilian R&D for long-term growth. This must be coupled with government action to change the attitude of the financial market towards companies who succeed in establishing a long-term civilian R&D investment culture, and to make long-term money more readily available, either through the markets or by direct investment.

In the case of defence diversification, government should bear an equitable and realistic share of the risk. The winning of MoD contracts should be dependent on defence companies developing diversification policies.

Maintaining the technological edge

The UK is also not sufficiently effective in turning its strength in science and technology into marketable products. Therefore, in addressing wealth-creation through diversification, the Defence Diversification Agency will need to consider not only future technologies and products, but also how the UK can improve the whole wealth-creation process, from science and invention to successful sales and well-satisfied customers.

The Foresight Panel identified key technology areas for future investment.[11] However, the process for prioritizing these technologies must be determined primarily by the customers (the MoD and industry) in the light of their specific needs, must be responsive to market pressures, and must take into account the extent to which UK industry can and should access technology from overseas sources, for example from European Union programmes.

The future demand for technology, in products and processes, requires that industry, government research establishments and universities work more closely and extensively together in high-quality, customer-led applied research programmes. To ensure that university applied research is better targeted at industry's non-military requirements and that industry has the commitment to exploit the technology, this category of work should primarily be co-funded by the UK Research Councils and the customer.

To create much greater partnership and joint commitment between industry, universities, Research Councils and the MoD, the Foresight Panel proposed two co-funded 'LINK'-type schemes. These would involve the aforementioned groups, and be supported by the DTI. The first scheme applies to the defence sector, and has the objective of encouraging truly dual-use technology programmes which maximize the value of the defence research funds. The second applies to the civilian aerospace sector.

The following figures are only *illustrative* of how these programmes might be funded. They are based on the existing joint grant schemes, in which the MoD and the Research Councils at present each contribute £5 million per annum. As proposed by the Foresight Panel, this would be augmented by the allocation of an additional £3 million from existing Research Council budgets, and a further £3 million from existing MoD funds. Industry and the DTI innovation budget would provide the balance of a proposed £24 million annual budget for the Defence Diversification Agency.

The Foresight Panel suggested that the £24 million annual funds for the development of dual-use technologies should come in equal £8 million shares from industry and the DTI, the MoD and the Research Councils. For the development of civilian aerospace technologies, the Foresight Panel suggests a total funding of £40 million. In its proposal, the Research Council's expenditure on civil aerospace would be co-ordinated and amount to some £20 million, matched by £20 million funding from industry and the DTI.

In each of these schemes, the Foresight Panel proposed that 50 per cent of the funds from industry should be eligible for 50 per cent matching by the Defence Diversification Agency, funded from the DTI's CARAD (Civil Aerospace Research and Development) and innovation budget.

These two schemes are currently being considered by the DTI and Office of Science and Technology (OST). The dual-use programme has made progress, and discussions have taken place between the OST and the European Union, which has expressed some reservations with regards to unfair competition resulting from such support for British industry. The progress of the civilian aerospace programme has been somewhat slower.

The Arms Conversion Project believes that these two schemes would help forge the links between UK industry and the Defence Diversification Agency.[12] Defence Diversification Agency would develop project proposals which would be judged for scientific quality and value to industry. A target should be set for a maximum period of six months between project submissions and release of funds. The ACP proposes that the Defence Diversification Agency should take responsibility for the management and progress of both dual-use and civilian aerospace programmes.

MoD procurement issues

There are various issues that govern the effective involvement of MoD procurement in a Defence Diversification Agency. The MoD has no formal responsibility to foster the strength of the defence industrial base or elastic creation. The openness of the UK defence market to foreign bidders is not reciprocated. The UK's leadership in the trend towards more off-the-shelf purchases poses real threats to the medium- to long-term capability of the UK defence technology base, and to the UK's ability to participate in international collaborative projects, and to compete in world markets.

There is a need to conserve the science and technology base and to exploit to the fullest possible extent its wealth-creation potential, while maintaining the right balance between off-the-shelf procurement and UK technology developments. The UK's difference in emphasis on 'offset' compared to other countries has placed it at a disadvantage in relation to its foreign competitors. There is a mismatch between industry's need for continuing activity to preserve skills and capability, and the increasing funding gaps and budget uncertainties inherent in defence programmes.

There is a need for procurement strategies and processes to align with the rapid evolution in technologies, such as information technology. Unlike its major foreign counterparts, the UK Ministry of Defence has no responsibilities to support development of improved business processes, especially in manufacturing. The current proposal says that the Defence Diversification Agency would be responsible for a review of MoD procurement policies and related defence science issues. This review would ensure that while there is continuing emphasis on value for money in the defence budget, there is also an emphasis on UK industry competitiveness and wealth-creation. Where possible, the MoD should ensure that its contracts are awarded only to companies which are prepared to develop appropriate diversification policies, and that agree to allow their management and trade union representatives to participate in the appropriate regional or national Defence Diversification Agencies.

THE UK DEFENCE AND AEROSPACE INDUSTRY

Financing issues for defence and aerospace

Financing issues are threatening the technology base and future viability of the defence and aerospace industries, which are key sectors of the economy. These issues need to be addressed in any national strategies for these industries. There is evidence that the problem is exacerbated in the UK by the financial pressures on companies during recent years, not least by the costs of restructuring and market pressures on prices, and by financial markets that systematically undervalue prospective income streams from long-term projects.

Significant success in civilian aerospace during the 1980s resulted from the successful government–industry partnership through the coupling of CARAD support for research and demonstration projects with well-targeted Launch-Aid support for initiation. The issue is how to develop this coupling further in the future. Because of the scale of the investment required, the long payback periods and the strategic importance of the industry, governments worldwide have heavily supported their defence and aerospace industries. In the defence and aerospace sector, R&D investment by the UK Government and industry is being reduced much faster than in competitor countries. If this reduction were being matched by an increase in civilian

R&D, it could be welcome. However, the reality is that this reduction threatens the UK technology base.

Technology demonstrators could provide a vital and cost-effective bridge between applied research and heavy development spending. But such demonstrators are critically underfunded today. Within technology demonstrator programmes, there is a need to address the area of business processes, given the increased emphasis on reducing cost and the time to bring a product to market. There is a need for a much stronger partnership between industry, universities, Research Councils and the MoD if the best use is to be made of the resources available for applied research.

Industry and government, in partnership, must respond to the threat of loss of market opportunity in the defence sector. They must redirect some of the current UK expenditure on defence research into civilian research, and must take steps to ensure that maximum spin-off or dual use is made of defence research. Technology demonstrator programmes should be the responsibility of the Defence Diversification Agency and be funded from MoD/DTI budgets and matching funding from industry. MoD-funded technology demonstrator programmes should become the responsibility of the Defence Diversification Agency, and consideration should be given to expanding their scope to include development processes as well as product technologies.

For civilian aerospace programmes, the terms of start up (Launch-Aid) support should not place UK industry at a disadvantage with respect to its competitors and 'partners'.

The Fourth Framework Programme

Stronger promotion of access to science- and technology-related funding available under the Fourth Framework Programme has been called for by the Office of Science and Technology in its 1994 publication *Forward Look*.[13] This consists of a £9.6 billion programme for research and technological development, and could be used for defence diversification projects.

A particular opportunity exists to link defence diversification and technology transfer initiatives occurring under the European Union Conversion Programme (KONVER).[14] Here, the Innovation Relay Centres would operate under the Fourth Framework Programme through the European Union. The role of the Relay network is to promote technological advance and competitive advantage. It should do this by linking, for technology transfer or matching purposes, institutions that develop research technology – whether industrial, academic or independent – with commercial production channels. Seven Innovation Relay Centres currently operate in the UK: one in Northern Ireland, one in Wales, one in Scotland and four in England. Potential also exists for forging better links between these centres and Technology Foresight initiatives, and with the defence spin-out (spin-off) and technology transfer programmes occurring through the activities of the MoD and DERA. However, so far there is a notable lack of co-operation between local authorities and the DTI/OST.

The missed opportunity offered by the Fourth Framework Programme could be recovered by a Defence Diversification Agency which included representatives of local authorities and government departments. This would allow both parties to benefit from the pooling of ideas and experience, and from contact with networks within which both sides operate.

DIVERSIFICATION PROGRAMMES

The Defence Evaluation and Research Agency

The UK Defence Evaluation and Research Agency is the largest research organization of its kind in Western Europe. It is a wholly-owned subsidiary of the UK Ministry of Defence. Its aim is to provide authoritative independent advice to the MoD and to ensure that the highest standards of quality and service are maintained in support of UK defence.

DERA has a total annual budget of £1,010 million, and employs some 10,948 people in support of UK defence. It is divided into four agencies:

1 The Defence Test and Evaluation Organization (DTEO) provides test facilities, employing some 3,255 employees with an annual budget of around £300 million.
2 The Defence Research Agency (DRA) has eight sectors, dealing with aircraft systems, human sciences, command and information systems, electronics, land systems, sea systems, structural materials and weapons systems. It employs 6,268 people, and has an annual budget of £600 million.
3 The Chemical and Biological Defence Establishment (CBDE) provides science-based protection services against chemical and biological warfare. CBDE employs 920 people, and has an annual budget of around £65 million.
4 The Centre for Defence Analysis (CDA) focuses on providing support to the policy and planning activities of the MoD. It employs 505 people, and has an annual budget of £45 million.

The Arms Conversion Project proposes that as part of a defence diversification strategy, the DTEO and the DRA should widen their scope to support the UK industrial base. In line with this wider remit, a new civil technologies sector could be created within the DRA. If ten per cent of the DRA's resources were allocated to this sector, then some £60 million and 600 staff could be employed in assisting defence diversification. Likewise, if ten per cent of the activities of the DTEO were earmarked for non-military use, some £30 million would be available for defence diversification activities. These proposals would result in a budget for defence diversification of £90 million per annum.

DERA also has an extramural research programme with a budget of £160 million per annum, £130 million of which goes to industry, and £30 million to academic institutions. The Defence Diversification Agency should propose ways of ensuring maximum civilian spin-off from this research programme.

The Pathfinder Programme

The Pathfinder Programme aims to find dual-use technology in private companies and universities.[15] Each year, around 300 proposals are submitted to the MoD, and about 25 per cent of these receive 50 per cent matching funding from the MoD. While there is no limit to the level of Pathfinder funding, most awards are for around £200,000. Around £15 million is awarded to Pathfinder projects per annum. DERA is looking at ways of increasing the overall Pathfinder spending, and placing greater emphasis on technology transfer between DERA and industry.

The Pathfinder Programme should be brought under the overall management of the Defence Diversification Agency's National Dual-Use Technology Research Centres. These 'centres' would more accurately be described as networks, each taking a different form. The Structural Materials Centre co-ordinates research. The Marine Technology Centre is based in Portsmouth. The Software Engineering Centre is a network. The Super Computing Centre supports computing needs. The Electronics Centre follows a portfolio approach. Additional centres under consideration are Telecoms, Information Systems, and Robotics. Each centre can give wide access to a network of researchers and research projects, and they develop projects which are subject to commercial investment appraisal.

It would appear reasonable that these National Dual-Use Technology Research Centres should be included within a defence diversification strategy. They are each managed by a board consisting of representatives from the major parties; this could be extended to include representatives from the Defence Diversification Agency.

Other possible Defence Diversification Agency contributions

The UK does not have access to a detailed analysis of the supply chains operating in the defence and aerospace industries. The Defence Diversification Agency could provide a range of programmes to assess the defence dependency of such companies, and to encourage diversification of their products and markets into non-defence-related activities.

The Sector Challenge is an initiative to promote the competitiveness of UK industry.[16] It seeks to bring together in one place the majority of government funding for sector-based activities. The overall purpose of Sector Challenge is to improve the competitiveness and long-term profitability of business sectors by promoting a culture of business excellence and innovation. The competition will be managed by the DTI, working closely with other government departments. The defence sector involves high levels of skills and technology, but lacks civilian market intelligence. Such funding could be used to provide improved civilian business performance or to increase business activity in emerging non-military industries.

Redeployment initiatives for defence industrial workers are vital. There have been substantial reductions in employment in UK defence industries. Over the period 1980–94, employment almost halved from 740,000 to 395,000. The *Redundant Defence Workers' Survey* of the Strathclyde Defence Industries Working Group revealed that less than half (48 per cent) of the former defence workers had found new jobs.[17] Furthermore, many of these new jobs were lower-paid and lower-skilled, and 28 per cent of them were of part-time or temporary nature. The survey also painted a worrying picture of the almost complete failure of employment support services to respond to such large-scale defence redundancies.

The proposed Defence Diversification Agency would be responsible for discussing long-term MoD procurement plans with defence industries and encouraging them to diversify. However, there remains the possibility of further job losses within the defence base. Some estimates predict a further 100,000 job losses in the next three years. The DDA should consider using MoD employment support services in defence industries, and should enter into discussions with the Employment Services and other government departments to ensure that maximum support is given to maintain high-tech skills within localities affected by defence industry reductions.

NON-INDUSTRIAL ACTIVITIES

The role of a Defence Diversification Agency would include not only industrial defence diversification, but also reuse of military bases, and redeployment of MoD personnel and civilian staff.

Re-use of military bases

The Defence Estate exists to provide operational bases, training areas, weapon areas, weapon ranges, accommodation and logistic facilities. It is regarded as a central asset available for the use of the whole MoD. The Secretary of State for Defence is responsible to Parliament for the control, use and development of the estate. The Defence Estate also provides income to the economy: £18.7 million in 1993–94. However, while this may appear substantial, it is a relatively small amount considering the size and nature of the estate.

While the overall size of the estate has changed little in the last two decades, it is now in the throes of major change.[18] The defence cuts already being implemented, and those being proposed by the Defence Costs Study, will lead to further reductions in the Defence Estate. Furthermore, the MoD is currently implementing a 'New Management Strategy' which includes effective management of the Defence Estate. The MoD expects to raise £600 million in the next ten years from land disposal receipts. Rationalization of land holdings and the early sale of surplus land and buildings is now regarded as a key priority.

The disposal process starts with the identification of potential surplus sites, either from estate rationalization studies, where the cost and feasibility of relocation is balanced against the potential sales value, or as a result of changing operational requirements or changes in military policy.

If no alternative defence use is identified for potential surplus sites, the MoD view is that in all cases, early contact with those local authorities with a direct or neighbouring interest is vital. The MoD departments have working parties in the areas most affected by changes in UK defence, and these are considered to be 'extremely useful in establishing mutual dialogue', since they allow full consultation with interested parties, and consideration of areas with mutual interest and possible mutual involvement.

The MoD staff in the Defence Lands Service must consider, with advice from external consultants where appropriate, the factors likely to determine the value of a site. These include: location, existing and potential use, legal restrictions, the presence of listed buildings and scheduled ancient monuments, and contamination, which can have important implications for the cost of future development. They must also consider the capacity of the local infrastructure to absorb additional development.

Negotiations with local planning authorities about planning consents to be implemented are normally undertaken by the MoD before a property is offered for sale. Indeed, the MoD is required to establish the development potential of all sites before they are offered for sale. In the current plan-led development environments, the MoD claims to seek to ensure that its land is zoned consistent with the plans of local authorities, to optimize the planning potential and hence receipts from disposals. The MoD considers the local plan process to be the proper mechanism for resolving planning and alternative use/development issues for all surplus defence sites. The MoD's stated aim is always to maintain a constructive working relationship with local planning authorities, and to welcome the opportunity, within the constraints of internal MoD procedures, to hold formal and informal discussions outside the statuary process.

It is normal practice to dispose of the freehold of surplus land in which there is no former owner or pre-emptive interest on the open market by competition, with the most favourable planning permission achievable, to maximize the return to the taxpayer. However, a range of other methods may be used if professional advice indicates that such alternatives would produce better value. It has been recommended that the Labour Party Strategic Defence Review should include a moratorium on the sell-off of the Defence Estate. No military base closures should happen until the end of this review. When military base closures or contractions/expansions are proposed, the details should be passed to the appropriate Regional Defence Diversification Agency (RDDA). The RDDA should investigate particular problems of released land, such as pollution, and it should also have responsibility for ensuring that local consultation on future use of such land takes place within national guidelines.

Redeployment of MoD personnel

In 1994–95, 2,440 Royal Navy, 7,015 Army and 1,640 Royal Air Force personnel left on redundancy terms. Further redundancies in 1995–97 involved up to 2,400 personnel in the Royal Navy, up to 400 in the Army, and some 8,600 in the Royal Air Force, and up to 17 Major-Generals were selected for redundancy during this period. Future staff requirements continue to be kept under review in the light of changing circumstances. Following these reductions in personnel, the armed forces are now experiencing difficulties in recruitment and retention.

Civilian staff numbers in the MoD have been reducing steadily since

1990. In addition, the 'Competing for Quality' objective and the implementation of efficiency measures under the 'Front Line First' programme will have a further impact on civilian staff numbers over the coming years. There have been over 10,600 redundancies since 1990, and a further 2,200 staff members were forced to leave before their planned retirement date. To assist military service personnel to gain access to civilian employment, the Ministry of Defence introduced a wide range of measures, including individual counselling and practical advice, briefings and training courses, a free and comprehensive job matching service, and information about job opportunities. These employment changes are so large in scale that it is recommended that a review of existing employment support measures, including discussions with appropriate civil service trade unions, should take place prior to any further redundancies being implemented.

DEFENCE DIVERSIFICATION AGENCY

The Defence Diversification Agency is being established by the MoD as part of its Defence Evaluation and Research Agency. The agency will work with a number of government departments, including the Department of Environment, the Department of Employment and Education, and Employment Services.

The following organizations are relevant to defence diversification.

- MoD
  - Department of Trade and Industry/Office of Science and Technology
  - Defence Evaluation and Research Agency
  - Department of Environment
  - Department of Employment and Education, Employment Services
  - devolved governments
- Regional/National
  - Research Councils
  - academia
  - industry associations
  - national trade unions
  - consultants
  - supervisory bodies
- Local
  - local authority organizations
  - local networks

- companies
- technology demonstration programmes
- consultants

The following structure has been proposed for the Defence Diversification Agency to ensure that it is both practical and responsive to the needs of local communities throughout the UK. It is hoped that the final structure of the DDA will incorporate as many of these suggestions as possible.

A Defence Council would be composed of appropriate Members of Parliament, together with representatives of the MoD and DTI, defence workers trade unions, industry associations, local authority organizations, academia and the Regional/National Defence Diversification Agencies (R/NDDAs).

An Executive Team would be funded by and based in the Office of Science and Technology. The team's resources should be supplemented by secondments from other organizations.

Eleven Regional/National Defence Diversification Agencies should be established. One should be in Scotland, one in Wales, one in Northern Ireland, and eight in England. Each of the R/NDDAs would establish its own Defence Diversification Councils. In Scotland, the Scottish DDA would work with the Scottish Parliament,[19] in Wales with the Welsh Assembly, in Northern Ireland with the Northern Ireland Assembly, and in England initially with the proposed Regional Chambers. It is envisaged that as England develops a series of Regional Assemblies, the Regional DDAs will become more accountable to these. Each of the R/NDDAs will establish Executive Teams. The one in Scotland will be based in Scottish Enterprise, in Wales in the Welsh Development Agency, in Northern Ireland in the Northern Ireland Office, and the eight in England in the Regional Government Offices.

Defence Diversification Council

The function of the Defence Diversification Council, with support from the Executive Team, would be to bring national actors in defence diversification together with representatives of R/NDDA, to liaise with the MoD, DTI and other appropriate government departments to establish the overall strategy of the DDA to mobilize resources and draw in and encourage participation, to co-ordinate activities undertaken by the R/NDDAs, including applications for funding from KONVER and other sources, and to conduct ongoing strategy development, monitoring and evaluation.

Executive Team

The function of the Executive Team would be to provide the Secretariat to the Defence Diversification Council, to implement the overall strategy of the DDA, as adopted by the council, to monitor the defence diversification initiatives and issue surrounding defence diversification, and to report on these to the Defence Diversification Council, to make recommendations to the council on overall policy and individual proposals, to assist R/NDDA in the development of regional/national initiatives within the overall DDA strategy, and to provide a day-to-day point of contact for those seeking information on the DDA and its activities.

R/NDDAs

The function of the Regional/National Defence Diversification Agencies would be to bring together the various regional/national actors in diversification to prepare regional/national diversification projects, to implement regional/national diversification initiatives under the overall DDA strategy, to monitor the impact of defence diversification initiatives, to report on regional/national initiatives to the UK Executive Team and the UK Defence Diversification Council, to send delegates to the UK Defence Diversification Council, and to make recommendations on the overall strategy of the DDA.

DDA proposed timetable

Immediately following the publication of the Green Paper on Defence Diversification, the Secretary of State for Defence should establish an initial Defence Diversification Council. This initial council would be based on the existing Civil Defence Forum, but expanded to include representatives of the Scottish, Welsh and Northern Ireland Office, defence trade unions, the Arms Conversion Project, the COST A10 programme, academia and other specialists in defence diversification issues.

The Office of Science and Technology would provide the initial secretariat to this group, and the first meeting would discuss the initial strategy of the Defence Diversification Agency. This initial meeting should be held within three months of the publication of the Green Paper.

THE ARMS CONVERSION PROJECT

The ACP was established by UK Nuclear Free Local Authorities in 1988 to help develop relevant and practical responses to the adverse effects on national and local economies resulting from the changing military and political situation after the Cold War.

The ACP organizes and participates in diversification and conversion initiatives, local networks, exchange trips, research conferences, seminars and workshops. It is a major voice in the issues of military base conversion and industrial and economic diversification, providing a positive and constructive input to local, national and international initiatives. It seeks to ensure the establishment of properly funded and resourced and highly accountable Defence Diversification Agencies to assist in the development and implementation of diversification initiatives.

The ACP works on the issue of diversification and conversion with local authorities, defence companies, trade unions, political parties, academics, peace groups, churches and individuals from across the globe. As part of its work, the ACP holds fringe meetings at various trade union and political party conferences, highlighting the need for a positive response from politicians and a participative role for trade unions and local authorities in the diversification process.

The ACP has attracted cross-party support for its activities. The project holds regular meetings with government departments, political parties, local authorities and trade unions, and is supported by local authorities, a wide range of national and local trade unions, peace groups and other organizations. The ACP is financed by contributing sponsors and subscribers to its publications. Financiers have a direct input into the ACP's work programme as well as direct access to the accumulated project expertise and information. The major sponsors of the ACP are local authorities. Trade unions, peace groups, church groups, political parties, non-governmental organizations and individuals may become subscribers to the ACP publications, gaining access to all studies undertaken and information collected.

REFERENCES

1 UK Ministry of Defence, *Statement on the Defence Estimates 1995* (London: HMSO, 1995).
2 See, for example, Defence Committee of the UK Ministry of Defence, *The Defence Estate* (London: HMSO, November 1994).
3 K. Hartley and Ines Newman (eds), *Defence, Research & Development and Diversification* (London: SEEDS, 1996).
4 Defence and Aerospace Panel of the Technology Foresight Programme, *Foresight*

Flyer (London: OST, Autumn 1996); Office of Science and Technology, *Foresight: First Progress Report 1996* (London: HMSO, 1996).

5   Reported in *Networker*, No. 23 (1988).

6   *Labour Party's Green Paper on Defence Diversification* (London: HMSO, 1998).

7   *Defence Diversification: Getting the Most Out of Defence Technology* (London: HMSO, March 1998). This builds on proposals put forward in the Labour Party's Green Paper.

8   K. Hartley and N. Hooper, *Study of the Value of the Defence Industry to the UK Economy* (York: University of York Publishing Co., 1996).

9   Department of Trade and Industry, *DTI statistics 1991* (London: HMSO, 1991).

10   *Table of UK Exporters, 1993* (London: HMSO, 1995).

11   Foresight Panel, *Foresight Flyer* (London: HMSO, 1995).

12   Arms Conversion Project, *Proposals for a Defence Diversification Agency* (Glasgow: ACP, 1997).

13   Office of Science and Technology, *Europe: Funding from the Fourth Framework Programme for Research and Technology Development* (London: HMSO, 1995).

14   Department of Trade and Industry, *UK KONVER Programme* (London, HMSO, 1995). KONVER is a European programme to assist areas in responding to the adverse effects on regional/territorial economics of the restructuring of the defence sector arising from reduced expenditure on defence and defence procurement, and in particular employment loss.

15   *Pathfinder* is the newsletter of DERA, based in Farnborough, England.

16   Department of Trade and Industry, *The Sector Challenge* (London: HMSO, 1995).

17   Arms Conversion Project, *Redundant Defence Workers' Survey* (Glasgow: ACP, 1996).

18   Private communication, 'The Closure of MoD Sites & MoD Disposal Policy: A Note by MoD for the Arms Conversion Project', June 1995.

19   Scottish Labour Party, *Labour's Contract for a New Scotland* (London: Labour Party, 1996).

4   GPS: Military Technology to Consumer Good

Dietrich Schroeer

INTRODUCTION

Many nations devote a significant proportion of their wealth to maintaining and improving their military capabilities. It is desirable to extract maximum civilian benefits from these military capabilities, particularly in the case of the military superpowers, the USA and the former Soviet Union, by converting military technologies to civilian use.

The Global Positioning System (GPS) is a military technology that has successfully penetrated the civilian market. This chapter examines why this particular technology transfer from military to civilian use was so successful. It first describes how GPS functions, its capabilities, and the technological elements that make it possible, then reviews the development history of GPS, and the extensions of its application from the strategic arena of missile-submarine navigation to use for navigation by civilian airliners to its incorporation into private recreational use.

Several features explain the successful civilianization of GPS:

1   the tremendous growth rate of the GPS technology, and its potential for continuing future growth
2   the existence of civilian interest in GPS applications even before the technology became available
3   recognition of the great benefits from GPS technology to both the military and civilian sectors, so that both sectors had a great interest in developing and improving it
4   experience of potential civilian uses of GPS on a personal level through its military applications.

These criteria seem to measure the likelihood of successful transfer of any technology from the military sector to civilian consumption.

THE TECHNOLOGY OF GPS

GPS was initially established for military purposes, yet it quickly became very attractive to civilian users. To understand the civilian interest in GPS, it is necessary to consider its technical history and how it functions. It is important to recognize that GPS is made up of technologies that were quite ready for exploitation, and that it has enormous potential for further improvement. When GPS was developed, the use of space for communication and navigation was well established, the electronic technology for transmitting precise signals was quite advanced, and the computer technology allowing ever smaller and cheaper receivers was available.

GPS's structure

The Global Positioning System is made up of three segments: the transmitter satellites, the operational control segment, and the passive receivers.[1] These three segments can be supplemented by ground- or space-based transmitters that permit differential location determinations (D-GPS). The GPS transmitting segment combines the concept of satellite-based transmission from the US Navy's *Transit* satellite navigation system, satellite-based atomic clocks developed by the Naval Research Laboratory as part of the *Timation* programme, and the pulse time-to-arrival technique developed by the Aerospace Corporation for the US Air Force as Project 621B.

In 1968, the US Department of Defense (DoD) established a tri-service steering committee to meld various proposed satellite navigation systems into one.[2] The *Navstar* GPS was accepted for deployment by the DoD late in 1973. The first *Navstar* satellite carrying an atomic clock for test purposes was launched in July 1974, the first operational GPS satellite for system-testing purposes was launched in February 1978, and the first GPS satellite carrying a nuclear detonation detection system was launched in April 1980. Contracts to develop user equipment were awarded in 1985. The first of the Block-II satellites for the operational GPS system was deployed in February 1989, and the GPS of 24 satellites was declared completely operational in December 1993.

The cost of the Global Positioning System in constant 1995 US dollars through to 2016 has been estimated at $8.5 billion for the satellite transmitters and $5.5 billion for military user equipment (consisting of about 160,000 receivers).[3] Including launching costs and nuclear explosion sensors, the total cost of GPS to the US military may be closer to $22 billion. This does not include civilian expenditure on receivers, which is rapidly overtaking military expenditure.

The GPS satellites

The US *Navstar* GPS transmitter network consists of six sets of four satellites, located in planes of 55-degree inclination, orbiting 20,051 km above the earth with a period of half a day. This configuration allows a user at any location on the earth to have at least four satellites in view at all times of the day.

Each current GPS satellite carries two rubidium and two caesium atomic clocks. These highly precise clocks control the transmission by each GPS satellite of a continuous radio signal carrying a pseudo-random noise code signal, which contains satellite position data and the exact time of the signal's transmission. The GPS satellites transmit information at two different frequencies in the electromagnetic spectrum, and in two different codes. The L_1 signal is at 1,575.42 MHz. It carries a Coarse Acquisition (C/A) code. This C/A code is known as the Standard Positioning Service (SPS), because it provides location information freely available to all users.

The L_2 electromagnetic signal is at 1,227.60 MHz. It carries more precise timing information for the Precision (P) service. The P service can be encrypted. The L_1 signal also carries a part of the P code, which is used in combination with the L_2 information to correct for ionospheric effects on the signal propagation. This combination of P signals, the decryption Y code and the ionospheric corrections is known as the Precise Positioning Service (PPS). Only authorized users have access to the decryption process.

The C/A signal had originally been intended to give a positioning accuracy significantly less than that provided by the P code. When this turned out not to be the case, the DoD decided to deliberately degrade the C/A signal by a process known as Selective Availability (SA). This degradation was first activated by the DoD in March 1990, then turned off again in August 1990 for the duration of the Gulf War, and finally reactivated in July 1991.

The next generation of GPS satellites will probably not incorporate this deliberate degradation. Instead, in time of crisis they will block the transmission and reception of public high-accuracy signals in areas where these would be useful to an enemy. There will probably be a second-frequency signal available on the open channel to allow civilian users to correct for atmospheric delays of the signals.

The operational control segment

The ground-based control segment of GPS consists of a master control station and five monitoring stations. The master control station maintains the master clocks for the whole system. The monitoring stations monitor the satellite transmissions for change in satellite orbits and satellite clocks. Based on these measurements, the control segment issues corrections at least once a day for the satellite clocks, and for the orbital parameters that each satellite transmits.

The receivers

A GPS receiver must determine the range to four satellites in order to calculate a three-dimensional fix of its own location. This range is determined by correlating the times of arrival of the signals from the four satellites. The time information from one satellite is used to set the clock in the GPS receiver. The times of arrival of the other three satellite transmissions are then used by the receiver to determine its position in three-dimensional space. If the receiver sees more than four satellites, its location can be determined with higher accuracy.

The receiver extracts its velocity from the variations in the time of arrival of the various signals. High-accuracy velocity measurements require frequent evaluations of the receiver location. The receiver can improve the precision of its location fix by integrating its location information over an extended period. But in this integration process, velocity information is lost. Such integration techniques are used in high-accuracy land surveys.

The GPS receiver also identifies the satellite transmitting each signal, to extract its orbital location. For critical applications in which no errors can be allowed, the RAIM (Receiver Autonomous Integrity Monitoring) system allows the receiver to evaluate whether the GPS satellite is functioning correctly. This is important for precision aircraft landings under conditions of zero visibility, where a malfunctioning satellite would lead to a crash.

In the higher-accuracy PPS system the GPS receiver compares the L_1 and L_2 signals received at different frequencies to account for signal travel delays induced by atmospheric effects. Hence, high-precision GPS receivers must be able to receive and analyse both the SPS and PPS signals. The military must therefore protect both its military and the civilian signals against jamming by an enemy.

D-GPS

GPS accuracies can be improved by Differential GPS processes. In D-GPS, a stationary ground-based receiver at a precisely known location is used as a reference point. It detects and then re-transmits GPS signals. These secondary signals all contain the same inaccuracies induced by the atmosphere and by the SA degradation. A GPS receiver compares the satellite signals it receives with those relayed by the differential station, to correct for such systematic inaccuracies. Of course, only those receivers in range of the relayed D-GPS signal can make these corrections.

A D-GPS relay beacon near an airport can greatly improve the bad-weather landing capabilities. Therefore, the US Federal Aviation Administration plans to establish a D-GPS Wide-Area Augmentation System (WAAS).[4] Similarly, the US Coast Guard is establishing a D-GPS network for navigation on coastal waters.

Geodesy

As the accuracy of GPS improves, it becomes critical to correlate this precise location information with equally precise maps. There is not much point in knowing where your boat is located with 10 m accuracy, if your navigation chart mislocates a reef by 100 m, with which your boat then collides. High-precision GPS information is opening up a large market for more accurate maps.

The capabilities of GPS

The accuracy of GPS's locating capability ranges from millimetres to 100 metres. The accuracy depends on whether the SA degradation is turned off, whether the direct signals are compared with signals from ground-based differential stations, and whether the measurement is long-term static or short-term dynamic.

The accuracy for GPS in the two horizontal dimensions is about 100 metres for the SPS, and about 16 metres for the PPS. These accuracies represent two standard deviations in the root-mean-square radial distance (or 2 drms). This means that 95 per cent of all location determinations are expected to fall within that distance from the actual location. Table 4.1 lists the sources of the various errors in the location determinations.

Table 4.1 shows that the error for the SPS measurements comes largely from the SA degradation. If that degradation were not imposed, the SPS precision would be 32 m, limited largely by timing errors introduced by the

Table 4.1   Sources of GPS positioning errors

Error source	Range error (2 drms)		
	SPS (SA on)	*SPS (SA off)*	*PPS*
Selective availability	48 m	0 m	0 m
Atmospheric delays	14 m	14 m	1 m
Clock and orbits	7 m	7 m	7 m
Signal reflections	2 m	2 m	3 m
Receiver noise	1 m	1 m	1 m
Total user equivalent range error	51 m	16 m	8 m
Typical horizontal dilution of precision (DOP)	×2	×2	×2
Total GPS horizontal accuracy (2 drms)	**101 m**	**32 m**	**16 m**

Notes: The user equivalent range error (UERE) is for a 2-drms uncertainty in two dimensions, 2 drms represents a circle about the true position containing approximately 95 per cent of the position determinations. An alternative derivation of this total error is described in Joseph L. Leva, Maartent Uijt de Gaag and Karen Van Dyke, 'Performance of standalone GPS', in Elliott D. Kaplan (ed.), *Understanding GPS: Principles and Applications* (Boston, MA: Artech House, 1996), p.262.

Source: National Research Council Committee on the Future of the Global Positioning System, *The Global Positioning System: A Share of National Assets – Recommendations for Technical Improvements and Enhancements* (Washington, DC: National Academy Press, 1995), pp.68 and 160.

atmosphere. The PPS eliminates that atmospheric error by comparing the time of arrival of two signals at different frequencies.

Although the deletion of the SA degradation will considerably enhance the civilian uses of GPS at no extra cost, expanded civilian use of the GPS system will come primarily through differential systems. D-GPS accuracies of 5 m should be readily achievable. For land surveys, accuracies of a few millimetres can be achieved. For military operations, the problems of the SA degradation is, of course, moot. However, in military operations it is desirable not to have to rely on ground-based differential techniques, hence such super-high accuracies are not likely for military operations.

PRE-GPS NAVIGATION AIDS

To understand why GPS is being so rapidly adopted for civilian purposes, we need to examine the predecessors to GPS. Earlier navigation systems not only led to GPS, they also generated civilian interests in such systems.[5]

The *Navstar* GPS grew out of the ground-based Loran-C and Omega navigation aid systems, and out of the satellite-based *Transit* navigation aid system. These systems were developed largely for military purposes, but together with the lower-accuracy civilian Loran-A system, they gave experience with navigation aids, and raised expectations of them. Table 4.2 lists the various navigational forerunners of the GPS system.

Loran-C

The first LOng RAnge Navigation (Loran) systems were developed by the USA during the Second World War as navigation aids for both ships and long-range strategic bombers. By the end of the war, 70 transmitters operating at 2 MHz were located globally, with 75,000 receivers on ships and planes. For example, a special temporary Loran transmitter chain was established to assist in the guidance of the strategic bombers that dropped the nuclear bombs on Hiroshima and Nagasaki. This system, known as Loran-A, became the standard system for surface ship navigation, both miliary and civilian. It was finally terminated in 1980.

After the Second World War, efforts were made to develop a longer-range navigation system using lower-frequency signals that bounce off the ionosphere. In 1955, the US Air Force decided to rely on inertial guidance systems for its strategic weaponry instead, so it was left to the US Navy to develop such a low-frequency system to assist the new SLBM (Submarine-Launched Ballistic Missile) submarines in their navigation. One such system was the land-based Loran-C.

Loran-C emits radio signals at 100 kHz. The signal itself contains pulses. From the pulse envelope, the receiver can extract an approximate location. From the differences in times of arrival (TOA) of the signals from different transmitters, the receiver can then determine its location. Although the Loran-C signal can travel only about 1,000 km, by 1974 eight per cent of the earth's surface was covered by Loran-C signals, most heavily concentrated in the North Atlantic. There was some additional coverage of the Pacific area for Vietnam bombing operations in 1971.

The location accuracy of Loran-C is limited by the consistency in the time-keeping among the various transmitters. Initially, each transmitter in the Loran system consisted of a master station controlling two slave

Table 4.2 Summary of electromagnetic navigation aids

System	Started	Transmission	Technique	Frequency	Wavelength	Accuracy	Application
Loran-A	1942	Land-based	Phase difference	180 kHz	[1.7 km]	>185 m	Naval
Transit	1964	Satellite	Frequency shift			185 m	SLBM
Omega	1966	Land-based	Phase difference	10 kHz	[30 km]	2 km	SLBM
Omega (Cs clock)	1983	Land-based	Phase difference	10 kHz	[30 km]	2 km	SLBM
Loran-C	1960	Land-based	Time to arrival	100 kHz	[3 km]	100 m	SLBM
Navstar	1990	Satellite	Time to arrival	1,575 MHz	[0.2 m]	16 m	All, global

Source: Owen Wilkes and Nils Petter Gleditsch, *Loran-C and Omega: A Study of the Military Importance of Radio Navigation Aids* (Oslo: Norwegian University Press, 1987), pp.22, 31–4, 73, 75, 99, 136, 138.

transmitters, and the receiver could only determine its position based on the two slaves in a single transmitter chain. By 1970, caesium-beam oscillators (atomic clocks) were available, so that the clocks of different chains could be synchronized. Thereafter, receivers could compare signals from different chains. Accuracies of 15 m could be achieved, and in 1977 an Army backpack receiver had a resolution of 65 m.[6]

Loran-C was developed to satisfy military requirements. It had no civilian applications until the 1970s, although such utility was claimed by the military whenever possible to help justify the systems, particularly when placing transmitters in foreign nations.[7] Loran-C was mainly a navigation aid for SLBM submarines, to help their missiles achieve satisfactory accuracies. In that case, the accuracy of the navigation aid must be good enough to target nuclear missiles, but the information must be received without revealing the location of the submarine. Loran-C had sufficient accuracy, but its signals could not penetrate water to a sufficient depth to allow the submarine to stay hidden by being submerged.

Civilian use of Loran-C was discouraged by the military.[8] Instead, the slower and less accurate Loran-A system was the mainstay for civilian ship navigation. It was only in the early 1970s that receivers had become sufficiently miniaturized and inexpensive to allow civilian uses. In 1973, 88 per cent of US merchant ships had Loran-A receivers, but only 38 per cent had Loran-C receivers.[9] However, when Loran-A was phased out, Loran-C became accepted as a general system. So, after about 1974, Loran-C underwent 'a transformation from almost purely military to a predominantly civil system'.[10]

The pre-GPS history shows that by the time the *Navstar* GPS was developed, civilian navigators had acquired experience with time-to-arrival navigation, with falling prices of equipment, and with civil exploitation of a military system.

Omega

A navigation aid system operating at a lower frequency is desirable for communicating with submerged submarines, since such electromagnetic radiation penetrates water to a greater depth. At the same time, the lower frequency means decreased accuracy for the location determinations. This is an unresolvable dilemma for SLBM submarines.

Omega transmitters emit their signals at low frequencies between 10 and 14 kHz. The receiver measures the travel time of signals not from pulse arrival times, but from phase shifts between the signals from the different transmitters. A precision of 2 km could be achieved for a wavelength of 30 km (10 kHz). Because these long-wavelength signals are reflected by the ionosphere, eight Omega transmitters were able to cover the whole globe.

Omega development began in 1958, perhaps initially for the Polaris submarines, but its accuracy of one km at a distance of 1,000 km was inadequate for SLBM submarine navigation. While the surface fleet did acquire Omega receivers, as did some Navy and Air Force surveillance and long-range transportation aircraft, they used it only as a system of last resort. Studies and experience showed that *Omega* was of little use as a navigation aid, and it provoked little civilian interest in GPS. Its low frequency, imposed for military reasons, inevitably degraded the accuracy of any location determination made by it.

Transit

Both Loran-C and Omega transmitted their signals from ground-based stations. Global coverage with ground-based transmitters required many stations, including some in foreign countries. The US satellite-based navigation aid system *Transit* was intended to satisfy that global requirement. In *Transit*, a receiver compared the frequencies of the signals received from several satellites. The Doppler frequency shift in these signals told the receiver its movement relative to the satellite, and this allowed a determination of the location of the receiver. Unfortunately, a location fix using *Transit* required that the satellite signal be measured for several minutes, and at very restricted times.

Transit became operational in 1964, and was made available for civilian use after 1967. By 1973, *Transit* could give 18–36 m accuracy for surveying a stationary location, but for moving objects, the frequency shift technique gave low accuracy. Hence, *Transit* was an inspiration for civilianizing GPS only in so far as it accustomed civilian users to the concept of using satellite signals for navigation.

DEVELOPMENT HISTORY OF GPS

Does the development history of the Global Positioning System reveal any reasons why it was so rapidly civilianized?

Factors that made GPS possible

The *Navstar* GPS was a logical development of existing navigation aids, but the earlier navigation systems did not have the required high accuracy and global coverage. Three major factors made GPS feasible:

1. the expansion in space capabilities, which allowed global coverage from satellites
2. atomic clocks, developed for use in space
3. the exponential growth in the capabilities of electronics.

Satellite

When GPS was planned in 1972, space capabilities were well developed. Humans had landed on the moon and returned safely to the earth, satellite communications technology was advancing rapidly, and the US Navy's space-based *Transit* navigation system was operational. Thus, a global navigation system based in earth orbit was clearly feasible.

Atomic clocks

As part of the *Timation* programme, atomic clocks had been successfully developed for installation in satellites. Consequently, the GPS programme could rely on clocks with the requisite time accuracy.

Electronics

Since the development of the transistor in the late 1940s, electronic technologies and computers have been growing exponentially in capability. Since GPS involves much computation, improvements in electronics led to improvements in the components of GPS, particularly in the receivers. Small, inexpensive hand-held receivers can now perform the same functions which earlier required large, expensive receivers.

Once GPS was deployed, more uses developed for it. This has led to increased demand for receivers. As more receivers have been produced, their cost has dropped, and they have become more capable. As the price of receivers has dropped, even more applications have become interesting, hence even more receivers have been purchased, and the price has dropped even further. The plot of the cost of GPS receivers as a function of the number produced is known as its 'learning curve'. This plot for GPS receivers has a shape typical of learning curves for technologies that have great potential for continuing growth. The GPS learning curve shows that an increase of a factor of two in sales has consistently led to a decrease of 23 per cent in the cost per receiver.[11]

Military demand for GPS

GPS for SLBM navigation

One might have expected the US Air Force to be very interested in a GPS navigation aid for its strategic bombers; but it opted for inertial-guidance navigation systems in order to avoid problems from electronic countermeasures by the Soviet Union. Therefore, the *Navstar* GPS was initially developed as a navigation aid for US SLBM submarines. Since their targets are located global distances away, in order to aim their missiles such submarines have to know their location accurately on a global scale, so ground-based navigation aids are inadequate.

The interest of the US Navy in GPS grew out of the usefulness it found in electronic navigation systems during the Second World War. After 1942, the Loran system was developed for more general use. By 1972, it was clear that considerable improvements could be made in the accuracy provided by such navigation systems. At the same time the accuracy of the SLBMs was also being improved, and the targeting strategies of the nuclear forces began to include tactical nuclear strikes. Improved accuracy in navigation was necessary to minimize collateral damage, and to ensure the destruction of deeply buried enemy missile silos and command bunkers with the smaller warheads planned for multiple-warhead nuclear missiles. GPS promised to provide more than enough accuracy for such requirements.

Other military navigational uses of GPS

A high-accuracy GPS capability would also be of benefit to military navigation in general. It was recognized early on that surface ships could certainly use such a global system. However, GPS was not initially seen as particularly useful for land operations. Its resolution did not seem high enough, and the receivers were initially too expensive for large-scale use by individual tank or troop units.

But now, GPS has developed primarily into a land navigation system. The Persian Gulf War of 1991 demonstrated its usefulness for global navigation, such as for *en route* navigation for naval vessels and military aircraft, but GPS came even more to be recognized as very useful for land-based navigation by tanks and infantry, helicopter search and rescue operations, and so on.[12] It was even used for munitions guidance in the Naval Stand-off Land Attack Missile.[13]

In fact, during the Gulf War, the reliance of the land forces on GPS became so great that the number of available P code military receivers was too small to meet the demand. The US Department of Defense purchased thousands of civilian receivers, and individual soldiers purchased sports GPS receivers on their credit cards and taped them to their trucks and tanks.[14] To allow the use of these civilian GPS receivers, the DoD turned off the SA signal degradation.

Since the Gulf War, there has been tremendous growth in military use of GPS receivers, both by the US Army and US Air Force. Besides the applications in navigation for military aviation and for land manoeuvres, GPS has much potential for navigation in special warfare, mine warfare, field artillery targeting, delivery of cargo by parachute, and search and rescue missions. The use of GPS for precision-guided munitions, including cruise missiles, will increase in future,[15] and GPS technology is being developed to allow the B-1 and B-2 bombers to carry out high-accuracy tactical bombing.[16]

The judgement of the National Research Council is that the precision GPS service 'meets most of the military's positioning and navigation accuracy requirements'.[17] But military uses of GPS are not increasing as rapidly as those in the civilian sector. The problem is that the military requirements are as concerned with security as with accuracy.

GPS to detect atmospheric nuclear tests

GPS solves one problem arising from monitoring above-ground nuclear tests.[18] In the 1963 Limited Nuclear Test Ban Treaty, the USA and the Soviet Union agreed to halt all nuclear tests in the atmosphere. The *Vela* satellite system was established by the USA to detect such nuclear explosions, and thereby monitor this atmospheric test ban. However, the *Vela* system, with its infrared detectors, could only *detect* the light flash from an above-ground nuclear explosion; it could not establish the *location* of the flash. In the case of the single South African nuclear test, the *Vela* satellites detected the light signature typical of a nuclear explosion, but could not locate the test.

The time of arrival of the light flash at various GPS satellites does reveal the location of the explosion with high accuracy. Such nuclear test monitors were installed in the GPS satellites after 1980.[19] This system can also locate and evaluate nuclear explosions in case of a nuclear war.

THE CIVILIANIZATION OF GPS

Miliary navigation systems were consistently presented to the US Congress with the argument that they would also aid civilian navigation, but the navigation systems Loran-C, *Omega* and *Transit*, were not used by civilians as much as they might have been. Civilian use of Loran-C did not begin until more than a decade after its establishment for SLBM submarines, but there was certainly the potential for large-scale civilian use of such high-accuracy systems. After the Second World War, the standard Loran-A system 'became a vital instrument for general navigation, civilian as well as military'.[20] A survey in 1955/56 showed that the standard Loran-A stations were used largely for civilian purposes.

However, the US Navy initially paid only lip service to the possibility of civilian use of the *Navstar* GPS.[21] The expansion of the *Navstar* GPS system into the civilian sector came about almost by accident. In September 1983, a Soviet fighter aircraft shot down a Korean airliner which had strayed into its territory as a result of navigational errors. In response to this incident, the US President, Ronald Reagan, offered to make GPS available for civilian use, directing the civilian US Department of Transportation to work together with the US Department of Defense to manage GPS.[22] In 1987, the US Coast Guard accepted responsibility for all civilian uses of GPS information.

Civilian applications of GPS came first in forestry and surveying. In 1984, civilian surveyors began to use GPS, enhancing its accuracy by differential methods. The initial mobile applications of GPS were in truck and ship navigation, while the aircraft navigation applications came more slowly.

Gradually, enough GPS systems were sold in the civilian market for the price of the units to decrease sufficiently to develop additional applications. In 1988, the first hand-held GPS receiver could be purchased for $3,500.[23] By the time of the Gulf War, hand-held GPS receivers cost $1000 or less.[24] After the Gulf War, veterans began to purchase them for recreational activities, such as hunting, fishing and hiking.

In September 1991, the USA guaranteed free public availability of the GPS Standard Positioning Service for at least ten years. In 1992, this offer was extended into the foreseeable future. In February 1994, the Federal Aviation Administration (FAA) approved GPS as a stand-alone navigation aid for all phases of flight, including non-precision landing approaches. The FAA cancelled the development of microwave landing systems for low-visibility landings and in June 1994 announced the implementation of the WAAS to improve GPS for civilian flight navigation use. In 1995, US President Bill Clinton reaffirmed the USA's commitment to provide GPS signals to the world civilian community. In 1996, it was announced that the SA degradation would be terminated by 2006 at the latest.[25]

As GPS receivers have become less expensive, and as their acceptance has grown, the major consumer market of the automobile industry has opened up. As a consequence of these new applications, civilian sales of GPS-related technologies have grown very rapidly: in 1993, $0.4 billion; in 1994, $0.8 billion; in 1995, $1.2 billion; in 1996, $1.9 billion; in 1997, $3 billion; in 1998, $4.7 billion; in 1999, $6.3 billion, and in 2000 (projected), $8.5 billion.[26]

Civilian usage of GPS now exceeds miliary usage, and non-aviation civilian usage is now the dominant portion of all GPS receiver sales. The accounting firm Booz-Allen & Hamilton has prepared an economic survey of the North American GPS markets.[27] They project that by 2003, the total number of GPS receivers will be 12.2 million, while the cumulative GPS market will total $42.3 billion.

Two major problems face the Global Positioning System in the future.[28] The first is the conflict between national security and civilian use of GPS. The second is the challenge of how to allow the military forces of the USA and its allies to use GPS while denying its use to enemies. The USA has recently made broad policy decisions about the future of GPS to solve both of these problems.

The USA has always assured civilian users that at least the low-accuracy signals will be totally open, but higher-precision signals are needed by civilian users, and denial of high-accuracy signals to enemies is becoming more difficult. The USA has therefore assured the civilian aviation community that not only will the GPS satellite system continue to be maintained for civilian use, but that the deliberate SA degradation will be terminated by 2006 at the latest. Furthermore, the next generation of GPS satellites will transmit an additional civilian signal that allows corrections for atmospheric accuracy degradation. Together, these improvements should increase civilian precision from the current 100 m to 3 m.[29] In addition, the US Department of Defense has given up its long-standing opposition to enhancement of GPS accuracy by such global differential systems as the WAAS.[30] Instead, in an effort to deny enemies access to the high-precision features of GPS, the USA is proposing to develop blocking capabilities for selected regions of the world.[31]

Therefore, in the future, GPS will provide improved accuracy in the information obtained by civilian receivers from the satellite signals. Additional improvements in accuracy will come from ground- or even space-based supplemental differential GPS, such as the WAAS, which is particularly intended to provide landing guidance to civilian aircraft. Within the USA, the US Coast Guard proposes to establish a Nationwide Differential Global Positioning System (ND-GPS), consisting of ground-based stations to cover the entire lower 48 states, Hawaii and the southern part of Alaska. Accuracies of 1–5 m are expected.

Mapping, geodesy and surveying

One of the earliest applications of GPS was surveying and mapping. From the very beginning of GPS availability, differential GPS techniques could give millimetre-level accuracy over distances as large as hundreds of kilometres. With such accuracies, GPS can even be used to study the motion of bridges.

At the moment, only about 0.4 per cent of total revenues for geographic information systems, now exceeding $10 billion,[32] are derived from services provided by satellites. However, the current 15 per cent annual growth rate in the satellite service suggests it will reach $0.5 billion by 2005. Currently, long measurement times are required because of the SA degradation. Eliminating this degradation would reduce the costs for real-time dynamic surveying.

Commercial navigational GPS uses

The original justification for civilizing GPS was its use in global navigation on the seas and in the air. That goal has broadened to include the use of GPS for aircraft landing in bad weather. At present, GPS can be used only for non-precision approaches. Although the Europeans have been conducting D-GPS landing tests,[33] the bad-weather landing problem has been harder to solve than initially expected.

The service provided by GPS with the SA degradation cannot cater for the higher accuracies and reliabilities required for landings in poor visibility. But even if SA is turned off, bad-weather landings will require local differential GPS support.

The market for GPS receivers in automobile and light-truck navigation is large. In Japan, many hundreds of thousands of automobiles carry GPS-based navigation systems. In the USA, this market is also developing, although primarily in luxury cars.[34] The navigation benefits not only apply to general map reading, but also driving off-road, emergency locating services, shipment tracking and potential future traffic management. For much of this GPS usage, improvements of GPS accuracy by elimination of SA degradation would be highly desirable.

Medical navigational uses of GPS receivers include guidance of helicopters transporting patients,[35] and locating people making emergency phone calls from cellular telephones.[36] The tracking by police of probationers who are restricted to a certain area is under development.[37]

Personal navigational use

The personal experience many soldiers had with GPS receivers during the Gulf War led to the tremendous growth in recreational GPS usage thereafter. For these personal uses, an accuracy beyond that dictated by SA degradation would be very beneficial, and would expand usage greatly.[38] National coverage by D-GPS signals could solve that problem, but with higher cost and complexity. The Booz-Allen and Hamilton study indicates that turning off the SA degradation would increase the total North American GPS market from $42 billion to $64 billion by 2003.[39]

Russian competition

Russia is developing a positioning system to compete with the US *Navstar* GPS.[40] GLONASS (Global Navigation Satellite System) is very similar in design to the US system. Its accuracy in its standard mode is 50–70 m, although it offers somewhat higher accuracy at high latitudes, whereas GPS is more accurate nearer the equator. Future ground-based differential stations promise higher accuracies. A new generation of GLONASS-M satellites will offer a standard accuracy comparable to that of GPS with its SA degradation turned off. Combining the GPS and GLONASS signals improves accuracy by 20 per cent, and gives higher reliability. Future combinations of precision signals from the GPS and GLONASS may yield resolutions as good as 4 m, but such receivers will be considerably more expensive.[41]

The competition from GLONASS is one factor encouraging the USA to eliminate SA degradation. European plans to develop an EGNOS (European Geo-stationary Navigation Overlay Service) to supplement GPS and GLONASS are putting further pressure on the USA to civilianize its precision-GPS services.[42]

It is true that even with the SA degradation, the civilian exploitation of the GPS has followed a tremendous learning curve, but removal of SA degradation should increase the steepness of this learning curve even further. The estimate that elimination of SA degradation would increase the civilian market from $42 billion to $60 billion for the ten-year period from 1994 to 2003[43] has placed tremendous pressure on the Department of Defense to do so.

DISCUSSION OF GPS CIVILIANIZATION

The very rapid increase in applications for GPS reflects the very successful transfer of a technology from military to civilian usage. While GPS began as a military technology, its civilian applications have grown so rapidly that now they far exceed the military ones. There are several factors that made this technology transfer so successful.

The great potential of GPS for technological growth

The civilianization of GPS has been successful because GPS technologies have improved at a tremendous rate, and continue to have much potential for future improvement. The technologies of satellite communications and the GPS receivers have been continually miniaturized and have became ever less expensive, as well as more reliable, and many of these improvements were made directly for the civilian sector, which meant that few restrictions could be imposed by the military.

It is tempting to label GPS technology a technological imperative. An accuracy of 15 m was not really necessary for SLBM submarine navigation, since the missile accuracy was 100 m or worse. But an accuracy of 15 m was relatively easy to achieve. Hence, this high accuracy was inevitably incorporated into GPS, and now the civilian sector is demanding access to this highest level of performance.

Prior civilian interest in GPS existed

The GPS technology transfer has been so successful because interest in its capabilities existed even before it became technically feasible. The earlier electronic positioning systems such as Loran-C had already found some applications in civilian navigation. While the resolution of these early systems was not good enough to extend these applications beyond navigation by ships, the potential usefulness of a global navigation aid had been recognized even before GPS was developed. Potential GPS users had experience with earlier navigation aids, hence latent demand for a global navigation system had built up, only requiring higher accuracy, lower costs and fewer restrictions imposed by the military to release it.

GPS technology is very useful

The transfer of GPS technology to civilian applications was so successful because it offered great benefits to both military and civilian users. A space-based locating system offered great benefits in military operations, and improvements in submarine navigation and in the locating of nuclear tests were very attractive to the military, so the military provided the financial support to develop, operate and improve the high-precision, space-based GPS.

At the same time, GPS also provided obvious benefits to civilian operators. Initially, these benefits related to the prior demand for global navigation, where even the relatively expensive early applications were quite useful, but the additional civilian benefits continued to develop very rapidly as the entrepreneurial civilian sector developed improved receivers for uses far beyond the initial interests – such as locating favourite fishing holes and manoeuvring automobiles.

This attractiveness of GPS technology is illustrated by its production learning curve. The more GPS technology has been applied in the civilian sector, the less expensive it has become. This price reduction has led to further increases in civilian usage. The result has been an almost irresistable pressure to make the full capabilities of GPS available to the consumer market by turning off the SA signal degradation.

Civilian experience with GPS

The transfer of a military technology to the civilian sector appears to be most successful when the technology has a direct impact on individual consumers. The initial GPS applications were in relatively impersonal uses by surveyors and for the navigation of ships and trucks, but the civilianization of GPS grew more rapidly when the experience with it became more personal for soldiers during the Gulf War of 1991. Experience with the unrestricted-accuracy GPS made many future civilian users aware of the ultimate potential of the unrestricted system, and demonstrated the usefulness of the technology on a personal level. While these experiences were not necessary for the ultimate civilianization of GPS, they certainly accelerated the process.

WHEN AND HOW CAN A MILITARY BE CIVILIANIZED?

Study of the Global Positioning System suggests that there are four major reasons for the successful transfer of the GPS technology to civilian applications:

1   GPS technology had and has tremendous growth potential.
2   GPS could satisfy a pre-existing civilian demand.
3   GPS technology offered great benefits to both military and civilian users.
4   Individual consumers had experience of the usefulness of the GPS.

The question is whether this successful civilianization of GPS provides a model for the transfer of other technologies from the military to the consumer. Do these four criteria form absolute requirements for any successful technology transfer? Let us consider two other major technologies which have previously been civilianized – one successfully and one unsuccessfully.

Jet aircraft: A civilianization success

Jet aircraft were developed by the military, first as fighters during the Second World War, and then as strategic bombers. In the USA, civilianization came about when the tanker aircraft for refuelling strategic bombers formed the basis for the Boeing 707 aircraft:

1   The jet engine had great potential for continuing improvements in power and fuel-efficiency.
2   The jet aircraft technology satisfied a pre-existing consumer demand. Passenger aircraft already existed, but did not have enough power and range to satisfy the intercontinental market. Jet aircraft for civilian passenger transport were recognized as desirable.
3   Significant improvements were made in jet engines, for both military and civilian aircraft. Consequently, the military was quite interested in supporting continuing development of that technology, leading to tremendous improvements in civilian passenger transport.
4   The aircraft industry produced both military and civilian aircraft, so such companies as Boeing experienced the benefits of those technological improvements, and could produce the civilian versions of the aircraft. Also, pilots of commercial aircraft had often flown equivalent military aircraft, and therefore had personal experience with the potential of that technology.

Given that jet aircraft satisfy the four civilianization criteria, it is not surprising that they have been successfully civilianized. In contrast, supersonic aircraft do not satisfy these criteria equally well, and expensive civilian transport has not been as successful.

Nuclear power: A civilianization failure

Nuclear power was developed for military applications in naval propulsion, for instance in missile-carrying submarines. Success in that area then encouraged the transfer of this technology to producing electricity in civilian nuclear electric power plants. The technology has been relatively unsuccessful in its transfer to civilian use, because several of the civilianization criteria were not met.

1   The technology of civilian nuclear power reactors has not developed as expected. The translation of naval nuclear reactors into civilian power reactors built by a large industry presented technical problems of scaling-up and maintaining safety standards. Local opposition and environmental problems led to increasing costs, so that the learning curve of civilian nuclear electric power reactors has been negative – the more reactors are produced, the larger the cost per megawatt-hour of energy produced.
2   At the time that nuclear power was transformed to the civilian sector, there was little demand for new energy sources, as oil was readily available and cheap. The Atomic Energy Commission heavily promoted the use of nuclear power by the electric power industry, trying to create an artificial demand through developmental subsidies.
3   There was great interest among the military in nuclear power for propulsion of naval vessels. Civilian industry was not as enthusiastic about nuclear power, and had to be persuaded through subsidies to 'go nuclear'.
4   The civilian nuclear power industry was misled by the great success of the military nuclear propulsion programme. When it became aware of at least some of the potential problems of nuclear electric power operations, it was persuaded by subsidies from government to continue to pursue this power source. The individual consumer had no direct experience with nuclear power, except the negative one through the experience of the Cold War.

Since the civilian nuclear electric power industry failed to satisfy the four civilianization criteria, it is no surprise that civilian nuclear power has been a relative failure.

CONCLUSION

The reasons for the successful transfer of GPS technology to civilian use seem clear. The four criteria for successful transfer – growth potential, prior demand, great benefits, and civilian awareness – were well satisfied by this technology. Other examples of civilianizing technologies suggest that to be transferred successfully to civilian applications, military technologies should meet these criteria as far as possible. In that sense, GPS technology provides a good example of what is required for a successful transfer of a technology from military to civilian applications.

REFERENCES

1   National Research Council Committee on the Future of the Global Positioning System, *The Global Positioning System: A Shared National Asset – Recommendations for Technical Improvements and Enhancements* (Washington, DC: National Academy Press, 1995), pp.150–62. Thomas A. Herring, 'The Global Positioning System', *Scientific American*, Vol. 274, No. 2 (February 1996), pp.44–50; p.47. Scott Pace et al., *The Global Positioning System: Assessing National Policies* (Santa Monica, CA: RAND, 1995), pp.260–1. Tom Logsdon, *Understanding the Navstar: GPS, GIS, and IVHS* (New York: Van Nostrand Reinhold, 2nd edn, 1995), pp.88–90.

2   Pace et al., *The Global Positioning System*, gives a useful chronology for the GPS on pp.262–3.

3   Ibid., pp.267–70.

4   NRC, *The Global Positioning System*, p.170.

5   Owen Wilkes and Nils Petter Gleditsch, *Loran-C and Omega: A Study of the Military Importance of Radio Navigation Aids* (Oslo: Norwegian University Press, 1987).

6   Ibid., p.101.

7   Ibid. On p.34 and pp.195–9, the authors describe the claims of civilian benefits made by the US military when trying to persuade Norway to accept a Loran-C station on its territory.

8   Ibid., p.205.

9   Ibid., p.208.

10 Ibid., p.209.

11 NRC, *The Global Positioning System*, pp.186–93.

12 Bruce D. Nordwall, 'Imagination only limit to military, commercial applications for GPS', *Aviation Week & Space Technology*, Vol. 135, No. 15 (14 October 1991), pp.60–1. James Gleick, 'Lost in space', *New York Times Magazine*, 26 October 1997, p.24.

13 NRC, *The Global Positioning System*, p.26.

14 NRC, *The Global Positioning System*, p.21. Nordwall, 'Imagination only limit', pp.60–1: 'In fact, the GPS receiver capability was so highly sought after that many Army units acquired commercial units GPS receivers for ground navigation and installed them with a variety of "field expedient" methods ... Most common was a Trimble Trimpac GPS receiver held in place in an M-1 or A-1 tank with duct tape that the troops called 100 mph tape.'

15 'HARM missile to use GPS, inertial guidance', *Aviation Week & Space Technology*, Vol. 147, No. 17 (27 October 1997), p.74.

16 William B. Scott, 'B-2 drops GPS-guided "Bunker-Buster"', *Aviation Week & Space Technology*, Vol. 146, No. 17 (21 April 1997), p.64.

17 NRC, *The Global Positioning System*, p.26.

18 Wilkes and Gleditsch, *Loran-C and Omega*, p.176.

19 Pace et al., *The Global Positioning System*, p.263.

20 Wilkes and Gleditsch, *Loran-C and Omega*, pp.32 and 64.

21 Pace, et al., *The Global Positioning System*, pp.263–6 carries useful chronology of GPS's history, including the development of civilian applications.

22 John Markoff, 'Finding profit in aiding the lost: a civilian industry is built on the military's locator technology', *New York Times*, 5 March 1996, p.D-1.

23 Bruce D. Nordwall, 'GPS success sparks new concerns for users', *Aviation Week & Space Technology*, Vol. 147, No. 22 (1 December 1997), pp.58–60.

24 Markoff, 'Finding profit'.

25 Philip J. Klass, 'GPS plan to add new security techniques', *Aviation Week & Space Technology*, Vol. 144, No. 15 (8 April 1996), p.59.

26 Bruce D. Nordwall, 'GPS technology ripens for consumer market', *Aviation Week & Space Technology*, Vol. 143, No. 15 (9 October 1995), pp.50–1. Nordwall, 'GPS success sparks new concerns'.

27 'North American GPS markets: Analysis of SA and other policy alternatives', Appendix E in NRC, *The Global Positioning System*, pp.180–200.

28 NRC, *The Global Positioning System*, pp.162–76.

29 'U.S. to expand civil GPS', *Aviation Week & Space Technology*, Vol. 148, No. 14 (6 April 1998), p.60. Matthew L. Wald, 'U.S. to improve satellite navigation system', *New York Times*, 30 March 1998, p.A-12. Bruce D. Nordwall, 'GPS improvements need quick decision', *Aviation Week & Space Technology*, Vol. 147, No. 22 (1 December 1997), pp.62–5.

30 Philip J. Klass, 'New GPS policy attempts to resolve key user issues', *Aviation Week & Space Technology*, Vol. 146, No. 24 (9 June 1997), pp.42–3.

31 Philip J. Klass, 'GPS plan to add new security techniques', *Aviation Week & Space Technology*, Vol. 144, No. 15 (8 April 1996), p.59.

32 Logsdon, *Understanding the Navstar*, pp.242–3.

33 Matthew L. Wald, 'Airlines announce plan to install improved warning system', *New York Times*, 16 December 1997, p.A-18.

34 Stephen Manes, 'Where exactly are we? Navigation that is hardly celestial', *New York Times*, 19 August 1997, p.C-5. Andrew Pollack, 'Computers let Japan's drivers play in traffic', *New York Times*, 6 October 1997, p.D-1.

35 'GPS goes to the hospital', *Aviation Week & Space Technology*, Vol. 145, No. 17 (21 October 1996), p.13.

36 Teresa Riordan, 'New device helps emergency workers find people who use a mobile phone to call for help', *New York Times*, 10 November 1997, p.D-2.

37 'A watch-sized locking bracelet', *Aviation Week & Space Technology*, Vol. 144, No. 19 (6 May 1996), p.13.

38 NRC, *The Global Positioning System*, p.38.

39 Ibid., pp.197 and 198.

40 Bruce D. Nordwall, 'Optimism grows for GPS/Glonass', *Aviation Week & Space Technology*, Vol. 145, No. 16 (14 October 1996), pp.58–9.

41 John Markoff, 'New product to improve satellite navigation', *New York Times*, 22 May 1996, p.D-5.

42 Bruce D. Nordwall, 'World pressure grows for regional GPS augmentations', *Aviation Week & Space Technology*, Vol. 147, No. 22 (1 December 1997), pp.66–8.

43 NRC, *The Global Positioning System*, op. cit., compare pp.193 and 197.

5 Nuclear Power in Space: A Dual-use Conflict

Paolo Farinella, Luciano Anselmo and Bruno Bertotti

INTRODUCTION

This chapter summarizes the history and problems of nuclear energy in space. Half a century ago, at the time of the 'Atoms for Peace' initiative, it was commonplace to believe that nuclear energy would not only provide energy 'too cheap to meter' on the Earth's surface, but also would become the key technique allowing humanity to travel in the solar system and establish colonies on the Moon and the planets. Science-fiction novels abounded with detailed blueprints for giant, nuclear-propelled spacecraft, and even serious science popularizers often discussed how the nuclear reactor engines of submarines such as the *Nautilus* would soon be adapted to the requirements of interplanetary spacecraft. As we shall see, these predictions were completely wrong, and the applications of nuclear energy in space, while important, have been much more limited.

A discussion of nuclear energy in space is interesting for several reasons. First, nuclear energy in space represents a clear instance of dual-use technology. Military and civilian applications have long coexisted for this technology, and have influenced each other in a number of ways. However, eventually the viable technologies for the two applications diverged, and nowadays the space nuclear power devices which are used or developed for military purposes have little in common with civilian ones. It is interesting to consider whether this is a unique development, or whether this is a specific example of a more general rule which holds whenever there are conflicting aims and methodologies of the two kinds of applications.

Second, as with many other technologies of the twentieth century, a number of unpredicted problems and hazards have come to be recognized in this case, and have triggered a lively debate in the scientific and technical communities, and in public opinion. The main concerns have been the potential re-entry into the Earth's atmosphere of radioactive material, the interference with space research into high-energy astrophysics, the potentially destabilizing military applications, and the hazards related to the production of space debris and subsequent collisions in Earth orbit. Only recently has it been realized that most nuclear reactors in space have been disposed at precisely the altitudes (700–1,050 km) where drag-induced orbit decay is negligible, but where the probability of collision with a piece of space debris is significant and is going to increase in future. The accidental impact break-up of a nuclear reactor, or of a future nuclear-electric propulsion spacecraft, may thus pollute a large shell of circumterrestrial space. These concerns prompted the United Nations General Assembly to approve a resolution in 1992 establishing criteria for the safe use of nuclear power sources in space. However, these criteria are restricted to non-propulsive systems and to current types of technologies and missions, so a consensus at the international level cannot be taken for granted in the long term.

Moreover, as often happens in such cases, these problems and dangers of nuclear energy have been badly misrepresented in the media, and either played down or exaggerated for political reasons by the parties involved. It is also interesting to note that some of the problems are of an environmental type, and despite the lack of biological ecosystems in space, it is difficult to make reliable predictions about future developments or to take suitable mitigation measures.

Third, it can be argued that in this case, as in many others, a Manichean, black-and-white approach is not going to work. While a comprehensive ban on space nuclear systems in general appears neither feasible nor desirable, additional 'rules of the road' are needed to address current and future safety concerns. In this context, it appears important to make a clear distinction between nuclear systems operating permanently in low Earth orbits, and systems launched from or assembled near the Earth but intended to operate in deep interplanetary space. While the former systems should be forbidden up to some maximum height that takes into account the collision hazard, the latter ones may be allowed as long as suitable safety measures or devices are put in place. Technologies ensuring reliable verification of obedience to such rules are currently available, and have been widely discussed in the open literature.[1]

A BRIEF HISTORY OF NUCLEAR ENERGY IN SPACE

It has been clear since the 1950s that the exploration of remote solar system bodies or hostile planetary surfaces, as well as the exploitation of space resources, will probably require nuclear power sources. In fact, extrapolating existing technologies, it is difficult to envisage any other reliable system able to provide electrical power of the order of hundreds or thousands of kilowatts for long enough periods. Far away from the Sun, where solar panels are ineffective due to the inverse-square relationship between light intensity and distance, nuclear power sources are essentially the only way to obtain energy. Moreover, no power-generating technique shares the other advantages inherent in the nuclear sources: compactness, robustness, ability to work anywhere in the solar system, resistance to harsh environments, and a high degree of system autonomy.

Since the beginning of the nuclear age, nuclear power for space applications has been considered as an energy source for both power generation and propulsion.[2] In the early 1960s, the *Orion* project was intended to use sequentially detonated nuclear charges to lift and propel a large spaceship. From 1955 to 1973, the USA was engaged in an extensive research programme *Rover* to develop solid-core nuclear rockets. Such rocket motors were predicted to provide as much as twice the specific impulse of the best chemical rockets, and were seen as the only realistic propulsion system for a manned mission to Mars (a view still held by many people today). Probably the best-known project was the so-called *Nerva* motor, intended to replace the J-2 chemical engines burning liquid hydrogen and oxygen in the upper stages of the *Saturn V* moon rocket. The suspension of *Saturn V* production after 1969, the rapid change in the political climate and shifting US national priorities led to the termination of the *Rover* programme in 1973. However, the research and development efforts resulted in a technical success, and could be the basis of a renewed interest in nuclear rocket propulsion at the beginning of the twenty-first century.

In the 1980s, the US Strategic Defense Initiative Organization launched a secret effort, code-named *Timberwind*, to develop a nuclear upper missile stage that could sharply increase the lifting power of military boosters, but the programme was abandoned in 1993. In June 1991, it was disclosed that the former Soviet Union had also been involved for thirty years in the development of a powerful nuclear-thermal rocket engine to support human missions to Mars. The experimental results had been good, but no application is foreseen at present.

Nuclear power sources to provide electrical energy for the functioning of a spacecraft have been more successful. Both the USA (since 1961) and the Soviet Union (since 1965) have used several nuclear power systems on board spacecraft. The thermal energy liberated by nuclear processes, such as the decay of radioisotopes or the controlled fission of heavy nuclei in a reactor, may be used directly for on-board thermal control, or indirectly for conversion into electrical power. The USA has launched 25 spacecraft equipped with radioisotope thermo-electric generators (RTGs), one probe (the Mars *Pathfinder*) with 6 grams of plutonium-238 oxide for thermal control on the Mars surface, and one satellite powered by a thermo-electric nuclear-fission reactor. The Soviet Union, on the other hand, has launched only three spacecraft equipped with RTGs and two lunar rovers with radio-isotope thermal generators, but has launched at least 36 satellites carrying on-board nuclear-fission reactors, producing electricity either by thermo-electric or thermionic processes.

In Earth orbit, the USA has used RTG power sources on navigational (6), meteorological (2) and communications (2) satellites. The last RTG launch into low orbit (mean altitude less than 2,000 km) took place in 1972, while two spacecraft in geosynchronous orbit were launched with identical RTG power sources in 1976. All the other plutonium-carrying devices equipped lunar (6), Martian (3) or interplanetary (7) missions. *Ulysses*, an ESA/ NASA (European Space Agency/National Aeronautics and Space Agency) probe, was launched in 1990, followed by the *Cassini* Saturn orbiter in October 1997.

The only nuclear reactor launched into Earth orbit by the USA, on board the spacecraft *Snapshot* on April 1965, was experimental in nature, and was not followed by other flights. This reactor, SNAP-10A, placed into an orbit with a 4,000-year lifetime, operated successfully for 43 days, until a series of spurious electronic signals shut it down. Owing to the safety design guidelines, it was impossible to reactivate the reactor from the ground, and it is now definitely quiescent. A hundred years after the launch, the radioactivity level in the core will be less than 0.1 curie, and when *Snapshot* finally re-enters the atmosphere, this level will be negligible. SNAP-10A was designed to disperse its contents in the upper atmospheric layers.[3]

On the other hand, since 1967 the Soviet Union has routinely operated thermo-electric nuclear-fission reactors in very low Earth orbits (altitudes less than 300 km). The *Romashka* reactors have been incorporated into at least 34 military spacecraft used for radar ocean surveillance of Western fleets. At the end of each mission, lasting typically a few months, the nuclear reactor was boosted into a 1,000-year lifetime orbit, and the core separated from the reactor housing at that high altitude; at least this has been the case in the flights following the *Cosmos* 954 accident. The Soviets also developed a much more sophisticated and capable thermionic nuclear-fission reactor known as *Topaz*. After many years of laboratory development, in 1987 Topaz was space-qualified in two space missions (*Cosmos* 1818 and 1867) in a 500-year lifetime orbit. In 1991, Russia offered the *Topaz* reactors to potential Western customers as part of a new effort to commercialize space technologies. In May 1992, two *Topaz* reactors were purchased by the USA for research and development purposes for $13 million. However, a subsequent plan by the USA to launch a *Topaz* 2 reactor to power an experimental electric propulsion system into space by the end of 1995 was cancelled.

As far as radioisotope thermo-electric generators are concerned, the former Soviet Union launched only two satellites equipped with RTGs in 1965, as part of two constellations of small tactical military communications spacecraft. Two more RTGs were used for thermal control purposes on board the Moon rovers *Lunakhod* 1 and 2.

In general, nuclear power generators in space, both RTGs and reactors, were initially developed for both military and civilian applications – that is, as dual-use technologies. Later on, the RTG technology found its main applications in civilian missions. Nuclear-fission reactor power sources were extensively developed by the Soviet Union, but these were not transferred to the civilian sector.

SAFETY CONCERNS

During the last decade there has been growing concern over the use of nuclear power sources in space. For instance, there have been proposals to ban nuclear power sources in Earth orbit. Even the launch of interplanetary spacecraft equipped with RTGs (in particular, the *Galileo*, *Ulysses* and *Cassini* probes) has been strongly opposed in the USA by some public interest organizations, for fear of accidents during the ascent or the near-Earth phases of the flight. Stronger protests and legal actions may be expected in the future.

Until now, the safety design criteria applied to the space nuclear power systems appear to have performed well in emergency situations. The first incident occurred on 21 April 1964, when the US Navy satellite *Transit* 5BN-3 failed to achieve orbital speed and re-entered the atmosphere. However, the SNAP-9A RTG performed as designed, burning up completely at an altitude of 45–60 km in the southern hemisphere. The second incident occurred on 18 May 1968. Owing to the international destruction of an erratic launch vehicle, the American *Nimbus* B1 meteorological satellite plunged into the Pacific Ocean about 5 km off the California coast. The SNAP-19 RTG carried on board was designed to re-enter intact, avoiding the dispersion of the fuel (Pu-238) into the environment. Five months later, the generator was recovered intact from the ocean floor at a depth of 90 m. The last incident involving a RTG happened in April 1970. The aborted manned Moon mission *Apollo* 13 carried in its lunar module, *Aquarius* a SNAP-27 fuel capsule to power the scientific instruments to be deployed on the lunar surface. Due to an oxygen tank explosion in the *Apollo* service module, the Moon landing was cancelled and the lunar module became a lifeboat for the crew, providing energy, oxygen and propulsion to the spaceship. When *Apollo* 13 approached the Earth for a safe splashdown, the lunar module was discarded and re-entered over the South Pacific Ocean at a speed of about 40,000 km/h. The fuel capsule was designed to survive the re-entry intact, and in fact the atmospheric monitoring of the impact area showed no release of plutonium-238 oxide. The SNAP-27 capsule probably lies on the floor of the Tonga Trench at a depth of more than 6 km. No adverse environmental effects have been reported up to now.[4]

The next accident involved the Soviet spacecraft *Cosmos* 954, a radar ocean reconnaissance satellite (RORSAT) equipped with a *Romashka* nuclear reactor. The standard procedure to boost the reactor into a safe orbit failed, and the satellite re-entered over Canada's Northwest Territories on 24 January 1978. Even though no large fuel particle was found in the subsequent field researches, several large fragments with high levels of radioactivity were recovered from the unpopulated area. Small particles of reactor fuel (uranium dioxide enriched to the 90 per cent level in U-235) were probably scattered over 100,000 square km with negligible impact on the environment. About 88 per cent of the fuel burned up during re-entry. Two further incidents involving Soviet RORSATs, *Cosmos* 1402 in 1983 and *Cosmos* 1900 in 1988, had less dramatic outcomes, due to design improvements of the safety systems. The reactor of *Cosmos* 1402 re-entered into the Atlantic Ocean, while that of *Cosmos* 1900 was finally boosted into a 'graveyard' orbit by a new safety mechanism.[5] In both cases, no release of radioactive material to the environment was detected.

As for the RTG-powered interplanetary probes, the only accident occurred when the Russian Mars 96 probe re-entered over the Pacific Ocean. Launched on 17 November 1996, this spacecraft was carrying six RTGs on board to produce electricity, plus an additional ten RTGs for thermal control, corresponding to a total of about 300 grams of plutonium-238 dioxide. Due to a malfunction in the *Proton* launcher's fourth stage, the probe ended up in an unstable Earth orbit and re-entered into the sea off the Chilean coast on 18 November 1996. In this case, as in all similar ones with modern-technology RTGs, no pollution of the environment by radioactive material has taken place, confirming that the safety measures were effective. The containers of the nuclear fuel are now lying on the bottom of the Pacific Ocean at a depth of about 6 km.

Is there any hazard from similar accidents in the future? Consider, for instance, the debate prior to the launch of the six-ton NASA probe *Cassini* towards Saturn. The hazard of concern was the potential for a release into the Earth's environment of some of the approximately 33 kg of plutonium dioxide in the RTGs. While plutonium-238 is not suitable for nuclear weapons, it is very dangerous to human health due to the alpha radiation it emits; but for toxic effects to take place, plutonium-238 must be reduced to very small particles and then delivered to the human body, and stay there over a long period. NASA claims that the ceramic form in which the plutonium dioxide is fabricated makes it highly insoluble in water, and limits the potential release of respirable particles.

The hazard with the *Cassini* mission is concentrated in two critical phases: the launch (by a *Titan* IV booster) with the Earth-orbit insertion phase, and the subsequent (August 1999) Earth swingby required to accelerate the probe towards Saturn. For the chances of a launch failure or of an inadvertent re-entry of the probe during the earth swingby, NASA has conducted an in-depth analysis, which incorporates human error and historical spacecraft reliability data. This analysis has determined a probability of about 1 in 1,000 for a launch accident with plutonium dioxide release, and about 1 in 1,000,000 for the Earth-swingby accident scenario.[6] Assuming that these estimates are reliable, it can be calculated that the corresponding risk factor (the probability of an accident multiplied by the number of estimated possible fatalities if the accident were to occur) is of the order of 100 fatalities. This risk is small compared to many other hazards, both natural and human-made, but is not entirely negligible. Some environmentalists have challenged the results of the NASA hazard study, recalling that similar optimistic estimates had been given prior to the 1986 *Challenger* space shuttle disaster. This debate is likely to continue in future, and it appears to be yet another instance of many such dilemmas arising in assessing the risk/benefit ratio for complex technological and scientific endeavours.

In summary, it appears that the greatest nuclear safety risk has come from military activities, particularly from Soviet nuclear reactors in orbit. However, the risk was greatest in earlier space activities, where the Cold War led to less concern being shown for safety.

THE SPACE DEBRIS THREAT

At present, the US Space Command tracks about 8,500 objects in Earth orbit with sizes typically larger than 20 cm.[7] About 21 per cent of them are inactive payloads, 17 per cent are spent rocket stages, 13 per cent are operational debris, and 43 per cent are fragments resulting from more than 150 unintentional or deliberate fragmentations. Only 6 per cent of the catalogued objects are active satellites and probes. More than 17,000 catalogued space objects have re-entered the atmosphere since the launch of the first artificial satellite in 1957. However, several investigations indicate the likely presence in space of 2,000 additional objects in the 10–20 cm size range, and some 70,000–150,000 objects in the 1–10 cm range.[8] These untrackable smaller particles constitute a growing hazard for space operations, mainly when large structures (for example, space stations) and manned vehicles (for example, space shuttle orbiters) are involved.

To make matters worse, any collision between orbiting debris may generate a cloud of many more objects, resulting in a significant increase in the collision probability for the remaining objects. This is a typical exponential growth process that may easily get out of control. It has been estimated that in the current situation, even with a zero launch rate of new spacecraft, the amount of space debris will continue to grow, eventually creating within a few centuries a 'debris belt' around the Earth (or more accurately, a 'swarm', as it will not be confined near the equatorial plane). Of course, this would make it impossible to carry out any prolonged activity in that region.[9]

Unfortunately, most nuclear reactors used in space have been disposed of in orbits with altitudes of 700–1,050 km, where the collision probability with a piece of space junk is largest. The break-up of a nuclear reactor could be triggered by a projectile of the order of only several centimetres in size. That break-up could then produce thousands of sizeable pieces of radioactive debris. Initially, these would be confined to a narrow 'ring' very close to the original orbit, but within a few years they would be dispersed by natural perturbing forces over a large volume of space. Such pollution will not be short-lived, since above 700 km height the atmospheric drag force is so small that the corresponding removal time for the fragments is of the order of several centuries.

An unpredicted recent development is the detection by radar observation of a previously unrecognized space debris family, with circular orbits of 600–1,000 km altitude and inclinations around 65 degrees.[10] There are some 60,000 such particles larger than 8 mm, and a few hundreds larger than 3 cm. The radar signatures of these objects are characteristic of conductive spheres, and their ballistic coefficients are consistent with mass densities of about 1 gram per cubic cm. Because of these orbital and physical properties, the particles have been identified as droplets of liquid sodium-potassium (NaK) coolant leaked from one or more of the nuclear reactor-powered RORSATs launched in the course of the military space programme of the former Soviet Union. After extensive analysis in both the US and Russia, the origin of this new source of debris has been traced back to the *Cosmos* 954 malfunction described earlier. To prevent the future occurrence of such accidents, the Soviet RORSATs were redesigned in such a way as to eject the fuel core from the reactor at the end of the mission. This would ensure the complete vaporization of the bare fuel core during an accidental re-entry into the Earth's atmosphere, as happened five years later with *Cosmos* 1402. However, the design change also affected the nominal mission profile. At the conclusion of each successful flight at low altitude, the nuclear reactor section was boosted up to the approximately 800 km 'graveyard' orbit, as before with *Cosmos* 954, but with the important difference that the fuel core was then ejected there. Unfortunately, the fuel core separation may well have been accompanied by a loss of sealing in the primary reactor coolant loop; the 13 kg of liquid NaK leaking from that loop is probably the detected debris.

Recent studies have shown that NaK droplets are not able to induce catastrophic fragmentation of large targets,[11] so they will not alter the future debris environment significantly as far as the possible onset of a collisional chain reaction is concerned. From this point of view, launch and debris mitigation policies, especially regarding satellite constellations in low orbit, play the dominant role in minimizing future collision problems. On the other hand, the NaK droplets may cause a significant and long-lasting increase in the rate of small-scale (non-disruptive) impact events in the 850–1,000 km altitude range, and they therefore give rise to an additional risk of serious damage for operational satellites orbiting there. This adverse effect will last for at least several decades.

Finally, a space debris swarm around the Earth is also a hazard for spaceships as they pass through it. Nuclear-electric propulsion (NEP) is being considered for Moon and Mars missions. Due to the low thrust-to-weight ratio, spacecraft equipped with NEP will travel along a spiral trajectory that very gradually expands until the escape velocity is finally achieved (weeks or months after the launch). Some preliminary estimates have shown that the NaK collision hazard may become significant for a large nuclear-powered spaceship after 2015, in particular if its spiral trajectory starts below an altitude of 1,100 km.[12]

In general, space debris is generated by both military and civilian activities in Earth orbit, and the debris affects both kinds of activities. However, the nuclear space debris comes more from military activities in space, while it may well have a greater impact on civilian space activities.

NUCLEAR WEAPONS AGAINST COMETS AND ASTEROIDS

An entirely different issue, concerning the possible use of nuclear weapons in space has been raised in the last decade. The Earth is sporadically hit by relatively large comets and asteroids (collectively called 'near-Earth objects', or NEOs). These hits may have dire consequences: impacts by NEOs about 10 km in diameter have been major agents in the evolution of the biosphere, triggering mass extinctions of living species such as the well-known Cretaceous–Tertiary boundary event of 65 million years ago, wiping out the dinosaurs. Even bodies 1–2 km in diameter may cause global climatic changes and other environmental effects that could pose a significant hazard to the future of civilianization.[13] Such impacts are expected to occur several times per million years.

It has been claimed that the only (or at least the most effective) technique to deal with an NEO that is on a collision course with the Earth, would be to deflect its trajectory by a powerful nuclear explosion nearby.[14] This suggestion has given rise to an intense debate involving many technical, political and legal aspects, as discussed, for example, by Gehrels, and Gerrard & Barber.[15] Some scientists have argued that a number of non-nuclear technologies can be developed instead to deflect threatening NEOs, especially if they were discovered and identified well in advance of the possible impact. But many nuclear weapons experts at both US and Russian military labs have enthusiastically supported the nuclear NEO-deflection scenario, suggesting that development and testing of the required technologies should be started as a high priority. This argument was also used for a time in 1995 by the Chinese delegation at the Comprehensive Nuclear Test Ban negotiations to argue that the test ban treaty should have allowed for some exceptions.

Despite the recent spate of Hollywood movies on this subject, however, it should be made clear to the public that the NEO-impact hazard is a very high-consequence, but a very low-probability scenario. On the one hand, the corresponding average risk factor, as defined earlier, is of the order of 1,000 fatalities per year, comparable, for example, to the casualty rate from airplane accidents,[16] but on the other hand it is extremely likely that no sizeable threatening NEO will be discovered in the future on a time scale of centuries. Under these conditions, the vast majority of non-military scientists and experts believe that the most prudent course of action is set up an intense observational effort to discover and survey the entire NEO population larger than, say, 1 km, which is estimated at about 1,500 objects. This can be achieved in a few decades, with modest investments of the order of $10 million per year. These observations would not be paralleled by preparing at the same time a NEO-directed nuclear-tipped missile arsenal, which could become dangerous in itself and hinder the nuclear disarmament process. If and when a really hazardous NEO is found in the future, the observations would hopefully ensure that there will be a sufficiently long warning time to develop and carry out appropriate measures to deal with the problem.

This debate is an example of military nuclear researchers and engineers trying to enter the civilian market. The objective for them is maintaining their jobs, and exploiting their expertise.

THE REGULATION OF NUCLEAR POWER SYSTEMS IN SPACE

The problems posed by both military and civilian nuclear power in space have led to proposals to ban all nuclear power systems in Earth orbit. Several reasons have been put forward so far to support such a ban on both civilian and military nuclear power systems in earth orbit:

1 No near-term civilian applications of such power systems are envisaged.
2 The highly radioactive core of activated nuclear reactors, as well as the toxicity of the plutonium used in RTGs, represents a potential hazard for the Earth environment in case of accidental re-entry into the atmosphere. This is true whether they are for civilian or military uses.
3 The collision of a small piece of artificial debris with a space nuclear power system could generate a cloud of radioactive fragments, soon dispersed by the perturbations over a large volume of space. This cloud would interfere with both military and civilian space operations.
4 The radiation (gamma rays and positrons) emitted by unshielded nuclear reactors in Earth orbit may 'blind' the instruments of space observatories devoted to research in gamma-ray astronomy, disrupting the study of this unique window on the most violent phenomena occurring in the universe. This is a civilian problem.
5 The possible military use of nuclear power systems in Earth orbit could stimulate an arms race. This is a military problem.

It is clear that a comprehensive ban on space nuclear systems could jeopardize important medium- and long-term civilian and military projects and possibly slow down the exploitation of a potentially promising technology. In 1992, the General Assembly of the United Nations approved a resolution entitled 'Principles Relevant to the Use of Nuclear Power Sources in Outer Space'.[17] This resolution established guidelines and criteria for the safer use of nuclear power sources in space. The endorsed set of principles 'applies to nuclear power sources in Outer Space' devoted to the generation of electric power on board space objects for non-propulsive purposes. This resolution is important because fills a gap in the international law on a critical topic. However, it only addresses nuclear power sources having 'characteristics generally comparable to those of systems used and mission performed at the time of the adoption of the Principles'. But much more capable systems could be developed in the near future by both the military and the civilian sectors, and the use of nuclear devices for propulsion is recommended for some new space missions. Therefore, some additional rules could be useful to address some of the safety concerns while taking into account the difference between a nuclear power system operating in a low Earth orbit and a system launched or assembled near the Earth to operate in interplanetary space or in an orbit far from the Earth.

Possible measures to allow some uses of nuclear power in space might include the following:

1   No nuclear power system should be operated in low Earth orbit, below some maximum height yet to be defined. In this region of space, only the *transit* of spacecraft carrying nuclear systems will be permitted.

2   Spacecraft carrying nuclear power systems could be assembled in low Earth orbit, provided their final destinations lie outside the forbidden region and an accidental release of radioactive material to the Earth environment will be prevented by safety mechanisms or procedures.

3   The orbits available for an extended stay and operation of space nuclear systems should lie at such altitudes that the interference with experiments dealing with gamma-ray astronomy would be reduced below a threshold yet to be fixed.

4   Nuclear devices used for propulsion might be activated in low Earth orbit only if the transit time is below a maximum yet to be set and if safety devices are in place to avoid any accidental contamination of the environment.

If such 'rules of the road' were implemented, the safety of nuclear systems operations in space and the confidence of the public at large on this issue would certainly increase. It would restrict military missions, but not unacceptably so, and it would allow most civilian projects currently envisaged. The possibility of carrying out interplanetary missions requiring nuclear systems would be preserved, as would be the option to assemble large nuclear spaceships in low Earth orbit and to develop new space nuclear technologies in higher and safer orbits.

REFERENCES

1   D.W. Hafemeister, 'Infrared monitoring of nuclear power in space', *Science and Global Security*, Vol. 1 (1990), pp.73–92.

2   J.A. Angelo and D. Buden, *Space Nuclear Power* (Malabar, FL: Orbit Book Co., 1985), pp.ix–xii.

3   Ibid., p.245.

4   Ibid.

5   L. Anselmo, A. Cardillo, A. Foni, A. Santoro and S. Trumpy, *Decadimento Orbitale e Rientro nell'Atmosfera del Cosmos 1900* ('Orbital Decay and Re-entry into the Atmosphere of Cosmos 1900') CNUCE Internal Report C89-05 (14 March 1989).

6   National Aeronautics and Space Administration, *Final Supplemental Environmental*

Impact Statement for the Cassini Mission (Washington, DC: US Government Printing Office, 1997).

7   National Research Council Committee on Space Debris, *Orbital Debris – A Technical Assessment* (Washington, DC: National Academy Press, 1995); pp.17–30.

8   Ibid.

9   D.J. Kessler and B.G. Cour-Palais, 'Collision frequency of artificial satellites: The creation of a debris belt', *Journal of Geophysical Research*, Vol. 83 (1979), pp.2,637–46. A. Rossi, A. Cordelli, P. Farinella and L. Anselmo, 'Collisional evolution of the Earth's orbital debris cloud', *Journal of Geophysical Research*, Vol. 99 (1994), pp.23, 195–210.

10   R. Sridharan, W. Beavers, R. Lambour, E.M. Gaposchkin, J. Kansky and E. Stansberry, 'Remote sensing and characterization of anomalous debris', *Proceedings of the 2nd European Conference on Space Debris, ESA SP-393* (1997), pp.261–9.

11   A. Rossi, C. Pardini, L. Anselmo, A. Cordelli and P. Farinella, 'Effects of the RORSAT NaK drops on the long term evolution of the space debris population', *Proceedings of the 48th International Astronautical Congress* (Turin, Italy, 6–10 October 1997).

12   J. Vedder, J. Tabor and D. Walyus, 'Orbital hazard for nuclear electric propulsion Earth-escape trajectories', *AIAA/AAS Astrodynamics Conference* (Hilton Head, SC, 10–12 August 1992).

13   C.R. Chapman and D. Morrison, 'Impacts on the Earth by asteroids and comets: Assessing the hazard', *Nature*, Vol. 367 (1994), pp.33–40.

14   T.J. Ahrens and A.W. Harris, 'Deflection and fragmentation of near-Earth asteroids', *Nature*, Vol. 360 (1992), pp.429–33.

15   T. Gehrels (ed.), *Hazards Due to Comets and Asteroids* (Tucson, AZ: University of Arizona Press, 1994). M.B. Gerrard and A.W. Barber, 'Asteroids and comets: US and international law and the lowest-probability, highest-consequence risk', *Environmental Journal*, Vol. 6 (1997), pp.4–49.

16   Chapman and Morrison, 'Impacts on the Earth'.

17   United Nations General Assembly, *Resolution 47/68* (14 December 1992).

6 Science and Nuclear Stockpile Stewardship

Dimitri Batani and Stefano Atzeni

INTRODUCTION

The Cold War is over, but it has left a huge legacy of weapons stockpiles of the nuclear superpowers. Even if no more warheads are produced, and even with the prospect of the eventual elimination of the existing ones, the safe disposal of the existing fissile material is a formidable task which will take decades. The current example of the disposal of Russian nuclear warheads, which is requiring the collaboration of the USA and is facing many problems, is enlightening in this context. The case of nuclear warhead disposal is an extreme example of problems posed by technology transfers: can nuclear weapons technology be safely eliminated without harming its civilian applications?

At the same time, even the recent long-awaited signature of the Comprehensive Test Ban Treaty (CTBT) seems to be creating a new additional problem. By prohibiting nuclear tests, the CTBT contributes to preventing the development of new nuclear weapons. But in the past, nuclear tests also served (or at least were claimed to be necessary) to periodically assess the reliability of stockpiled nuclear weapons. Opponents of the CTBT in the USA and elsewhere have been arguing that as time goes on, it will become increasingly difficult to guarantee nuclear deterrence with a much reduced number of nuclear weapons whose reliability is difficult to even assess.

In this context, the USA and France, and to a lesser extent other nuclear powers, have started complex programmes to replace nuclear tests with computer simulations and laboratory experiments. However, such programmes have raised concerns about their costs, their effectiveness, and above all, the possibility that they might lead to the development of new and more sophisticated nuclear weapons. Such developments might jeopardize the spirit of the CTBT, and give an advantage to those few countries which can afford them.

Assessing the safety (the 'health status') and the reliability of the existing nuclear stockpiles is a serious problem. It seems reasonable to discuss the size, cost and specific realizations of programme aiming at the above goals, while acknowledging arguments supporting their necessity. Given the global implications of nuclear weapons, this is not a matter of interest to nuclear powers alone – consider the consequences of a release of radioactive materials. Even more importantly, it is clear that the worries concerning the impossibility of maintaining nuclear deterrence will possibly become stronger and stronger as nuclear arsenals shrink and age. It would seem wise to give proper attention to such concerns, and find means to alleviate them, especially since the process towards nuclear stockpile reduction and elimination is not yet fully set, and could even be reversed.

The problems of nuclear material disposal and safeguarding are, of course, complex and multi-faceted. This chapter will sketch such stewardship programmes, and will also try to discuss and clarify some of the issues raised in this introduction. While doing so, it will be necessary to review how nuclear weapons work, what the traditional goals of experimental nuclear explosions were, and how and to what extent it is possible to simulate a nuclear explosion in the laboratory.

HOW DO NUCLEAR WEAPONS WORK?

The basic operation of nuclear weapons is described in various reference books.[1] Here, we will only review a few items which will be useful for the subsequent discussion.

Fission weapons

It is important to distinguish between fission weapons (the so-called 'atomic bombs') and fusion or thermonuclear weapons ('H-bombs'). In most atomic weapons, a certain amount of fissile material (highly enriched uranium or plutonium) is surrounded by a chemical explosive. The detonation of the latter causes a strong compression of the fissile core, so that a critical-mass condition is achieved and the nuclear chain reaction can start. The basic elements of the 'normal' atomic weapons are:

1   the fissile fuel and the critical-mass concept
2   the chemical explosives needed to drive the implosion of the nuclear fuel

3   the electronic and triggering system
4   the delivery system.

The physics of the critical-mass concept is very well known, and there is no major scientific challenge related to the development of atomic weapons. The challenges are rather the acquisition of the fissile nuclear fuel (a key aspect of proliferation, which is discussed in detail elsewhere in this book),[2] and the technical solutions needed to drive the fuel implosion (including the design of the conventional explosives and of the electronic and triggering systems). Also, the availability of a delivery system should be regarded as a critical issue, without which the nuclear explosive cannot be really considered a usable weapon.

Fusion weapons

Thermonuclear weapons release most of their energy by thermonuclear fusion reactions. The key concept here is the use of the energy released from a small fission device (the 'primary') to ignite a 'secondary' system, which contains the thermonuclear fuel. Part of the energy emitted from the primary, in the form of X-rays, is confined inside the external casing of the weapon and induces the compression and heating of the secondary, whose detonation is started by a complex mechanism which also uses some of the neutrons produced in the explosion of the primary.

A key issue for the realization of thermonuclear weapon is the availability of an atomic fission weapon. But unlike the case of fission weapons, the physics mechanisms of a fusion H-bomb are not known in detail, and the design of the weapon is rather complex. This means that the availability of the primary system and of the thermonuclear fuel for the secondary system is not enough to ensure the successful operation of the H-bomb design. According to some reports, this may have been the case of one of the recent Indian tests, which was claimed to be a thermonuclear weapon. The yield registered for that test by the international monitoring system was so low that it could correspond to the explosion of a primary system alone: the secondary system may have failed to ignite.[3]

THE RATIONALE OF NUCLEAR TESTS

The technical goals of nuclear tests can be summarized as follows:

1   to allow the development of new atomic weapons

2   to acquire data on how bombs work, and on how materials behave in the conditions of nuclear explosions, which are then useful for optimizing the weapon design by minimizing the use of fissile materials and/or to maximize some particular destructive effects

3   to 'qualify' the mass-produced bombs for military use; as in any other industrial product, at the end of the development phase even nuclear weapons enter a mass-production phase and are different in many respects from the 'hand-made' devices used to demonstrate that a new design is working

4   to periodically test the safety and reliability of previously qualified weapons; in very simple terms, it is important to be certain that a bomb will always explode when it is required to do so, and never explode when it should not![4]

The first two aspects are related to the design and optimization of weapons, the second one in particular to the acquisition of knowledge that can be used to refine computer simulation codes. In this respect, it will be recalled that the official justification presented to the French National Assembly for the last series of nuclear tests in Muroroa was the acquisition of more precise data than those already available, in order to optimize the computer codes used in numerical simulations of nuclear explosions.[5] However, at least one of the tests carried out in October 1995 had the goal of qualifying a new weapon – the model TN75, which has been deployed on French nuclear submarines since April 1996.[6]

However, the majority of nuclear tests have fallen into categories (3) and (4) above. Concerning the fourth category, it is important to recall the procedure used by the USA to verify the reliability of its nuclear weapons. A few randomly chosen weapons of each different model are completely dismantled and each component is checked. The non-nuclear parts are exploded and the device is rebuilt, replacing the parts which have a limited lifetime or have deteriorated. Finally, until recently a nuclear test was performed periodically, again with a randomly selected device. According to many in the weapons establishment, such tests were seen as giving the final assurance of the bomb's reliability.

However, it must be stressed that some prominent scientists who are experts in weapons programmes, such as the Nobel Laureate Hans Bethe or the former Livermore National Laboratory director Herbert York, claim that the reliability and safety of the arsenals can be guaranteed without testing,[7] although testing is essential for the development of new weapons. Indeed, uncertainties in physical data require testing to check that any newly designed thermonuclear weapon works properly. And perhaps more importantly, the military system would be very reluctant to introduce into its arsenal any weapon that has never been tested.[8] In the opinion of Bethe and York, a CTBT could present a serious obstacle to nuclear weapons proliferation without endangering weapons safety and functionality.

THE US AND FRENCH THERMONUCLEAR 'SIMULATION' PROGRAMME

In 1993, the US Clinton administration launched the Science Based Stockpile Stewardship (SBSS) programme, with the aim of guaranteeing the safety and functionality of the nuclear stockpiles even under a CTBT regime.[9] Following advice from governmental and independent committees, a programme with a cost of around $40 billion was defined.[10] France envisaged similar activities, and indeed, as mentioned above, the nuclear tests in Muroroa have been presented as a sort of preparation for this step.[11] Although the available information is limited, the UK and China also appear to be designing stewardship programmes.

In the USA, the establishment of such a programme has strongly contributed to removing the opposition by political conservatives to the US signature of the Comprehensive Test Ban Treaty.[12] On the other hand, its cost and possible proliferation implications have raised worries in parts of the US peace movement, which sometimes views stockpile stewardship only as instrumental to the signature of the CTBT, but without any specific merits in itself. It views it, above all, as a means for preserving the apparatus of the US national nuclear laboratories.[13] All these points will be discussed briefly in the following sections.

What is the basis of the stockpile stewardship and similar simulation programmes? As already indicated, many nuclear experts believe that it is now possible to acquire all the data needed to assure the health of the stockpile without performing any nuclear explosion tests. According to this view, tests can be replaced by the extensive use of computer simulations and by scaled-down laboratory experiments which do not involve any fissile materials. In particular, some aspects of thermonuclear weapons can be simulated through laser-based experiments. Such simulations would be necessary due to the unavoidable approximations or inaccuracies of the data on which the mathematical computer models are based, and due to the similarity between the physics of the studied phenomena.

For this reason, on 29 May 1997 the construction of the biggest laser system in the world, the National Ignition Facility (NIF), was begun at the Lawrence Livermore National Laboratory (LLNL) in the framework of the Science Based Stockpile Stewardship programme. NIF will be completed in 2002, at a cost of about $1.2 billion. According to a government leaflet, it will serve 'America's national security, energy, science and economic future'.[14] At the same time, in France, near Bordeaux, the Direction des Applicationes Militaires of the Commissariat à l'Energie Atomique (CEA), has started the construction of a twin laser, called *Megajoule*, which will be completed in 2010, and will cost 6.5 billion francs[15] (a smaller version will already be operational in the first years of the twenty-first century).

The debate over nuclear weapon test simulation had a major impact on French public opinion. For instance, in 1995, former president Valery Giscard d'Estaing, interviewed by the newspaper *Le Figaro*, stated that 'nuclear tests can nowadays be justified only if necessary to assure the passage to simulation'.[16] This was followed by an article with the title 'The time for simulation'.[17] In turn, the newspaper *Le Monde* published an article entitled 'France acquires a giant laser to simulate nuclear tests'.[18]

It must be said that there is an ambiguity in the formulation of the objectives of these programmes. The goals of the simulation programmes are in fact wider than weapon simulations. In general, they aim at the related goals of assuring the reliability of the nuclear stockpiles while maintaining scientific and technical communities with competencies in nuclear weapons. However, many feel that the preservation of the nuclear stockpile has little to do with the capability of fully simulating nuclear explosions. As has already been noted, the guarantee that a weapon will work is mainly related to the functioning of the primary system and its technological aspects. The relevant physical principles have been well known for a long time, hence the critical issues concern the chemical explosives, the electrical systems and the corrosion of the materials. In this respect, in a CTBT regime, only the problems of long-term ageing of the nuclear components of nuclear weapons could present aspects which have not been previously faced. The following describes how serious this point is. It is not clear that the solution of the ageing problems requires the capability of fully simulating explosions. Furthermore, the competencies required of the scientists and engineers for handling, assessing safety and reliability and dismantling nuclear bombs are perhaps not exactly the same as those required for designing nuclear weapons.

The preservation of the scientific knowledge about nuclear weapons systems concerns the whole physics of the system, and in particular, of the processes taking place in matter under the extreme conditions of temperature and pressure reached during the different phases of the explosion (including the ignition of the secondary system). This has given rise to much criticism, since performing stockpile stewardship activities would allow the USA to maintain a nuclear design capability.

Before discussing such issues, it is useful to sketch the simulation programmes. The US Science Based Stockpile Stewardship programme consists of the following five items:

1   hydrodynamics experiments
2   high-energy density experiments, in which pressures above 1,000,000 atmospheres are reached
3   computer modelling and numerical simulation
4   microfusion
5   study of non-nuclear aspects.[19]

Of the $40 billion SBSS total cost, about $2 billion is devoted to computer modelling and numerical simulation, through the Advanced Strategic Computing Initiative. About $4 billion are allocated for the construction of new facilities (although the NIF total estimated lifecycle cost will be $4.5 billion).[20] The vast majority of the stockpile stewardship funds are directed towards issues which are technological rather than scientific.

The hydrodynamics experiments consist of the chemical explosion of a primary system in which the fissile material has been replaced by another material of similar thermodynamic and mechanical properties – for example, natural uranium is suitable for this purpose, being essentially identical to enriched uranium-235, except for its nuclear properties. Such experiments are diagnosed by high-energy X-ray radiographs, which need special machines generating very intense, hard X-ray pulses. These are, for instance, the Dual Axis Radiographic Hydrodynamic Test (DARHT), which will operate near Los Alamos,[21] and the Accelerator for X-ray Imaging (AIRIX) under construction at Moronvilliers, in the French Marne Department.[22] The cost of the DARHT amounts to about $48 million, while the upgrade of PHERMEX (an existing facility at Los Alamos National Laboratory) will cost about $28 million. Finally, one other facility has been proposed, the Advanced Hydrotest Facility (AHF), with a cost of $422 million.[23]

It is important to note that the Comprehensive Test Ban Treaty prohibits so-called 'hydronuclear' experiments, in which a fissile primary is brought to the critical stage, but in which the chain reaction is prevented from diverging. On the other hand, the situation is not so clear concerning 'subcritical' tests – hydrodynamics experiments involving a small amount of fissile material that cannot reach the critical-mass condition even at high compression.[24] While a few experts considered them essential to the simulation programme, according to many others they do not add much to what can be learned from hydrodynamics experiments.[25] On the other hand, subcritical tests could threaten the CTBT, *de facto* changing it to a (Very) Low-Threshold Test Ban Treaty.

Critics say that even under the present CTBT regime, the task of verification would be very demanding, since it would be very difficult to distinguish a hydronuclear experiment from a hydrodynamics one. In this respect, what is really important is how much the side which violates the treaty could gain in knowledge compared to the risks it undertakes in violating it. In this context, not much can be gained from hydronuclear tests, and the political risks would be considerable. Hence a proposal has been made that hydrodynamic experiments should be performed above ground, in order to make verification easier.[26] This proposal does not seem particularly useful, and is also dangerous because of the possibility of pollution and accidents.

The high-energy density experiments provide information on matter under conditions comparable to those occurring in nuclear weapon explosions. In these experiments, a laser beam is focused on very small targets to study X-ray generation from hot matter, or to generate Megabar pressures and study the equation of state and the radiative properties of matter. Such experiments can certainly be realized with laser systems such as *Megajoule* or the National Ignition Facility, but the relatively smaller existing lasers (such as *Phebus* at the CEA Laboratory of Limeil, *Nova* at the Lawrence Livermore National Laboratory and *Omega* at the University of Rochester) have already proved to be adequate in many cases. They are used by civilian researchers for equation-of-state studies of interest to materials science and astrophysics.[27] Other high-energy density experiments are performed by using high-energy pulsed-power sources, which are less effective than lasers in generating high pressures in matter, but are capable of injecting, relatively cheaply, larger amounts of energy into matter contained in volumes of the order of a fraction of a cubic centimetre, which then act as an intense very high-temperature X-ray source.[28]

Computer simulation involves the numerical solution of the complex mathematical models of matter and radiation which describe a nuclear explosion. It is important to recall here that numerical simulation has always been used in the study of nuclear weapons and has often driven the development of novel computer systems and numerical analysis techniques. Despite order-of-magnitude progress in computing capabilities, present computer simulations are still far from the capability of fully simulating, with high resolution, all the aspects of the explosion of a real nuclear device.[29]

Microfusion is the aspect of stockpile stewardship which more directly concerns the giant lasers, such as *Megajoule* and the National Ignition Facility, and will be discussed in the next section. Finally, the study of non-nuclear aspects of nuclear weapons is of extreme importance. This implies periodic checking of all the components of a nuclear weapon, and of the launching, control and guidance components and so on, of the delivery systems. These can be classified as technological rather than scientific activities. According to the previous discussion, this is the part of the stockpile stewardship programme which is most relevant to the goal of assuring the safety and reliability of the nuclear stockpiles. Indeed, it is also the part which is taking the largest share of the money, for example the cost of new facilities such as DARTH and NIF, and of the strategic computing initiative, accounts for less than 15 per cent of the total SBSS cost.

It is worth observing that even if the US and French stewardship projects are very similar, the emphasis on the various parts and the relationship with the scientific community is rather different in the two countries. In the USA, the emphasis is also on the contribution to the research aimed at fusion energy production by inertial confinement fusion, and on the civilian and industrial use of lasers. A significant part of the NIF laser time will be devoted to civilian experiments, proposed and performed by university researchers. In the mind of the Clinton administration, this will also help to remove any doubt concerning the possible use of the National Ignition Facility and of the stewardship programme in general, for the realization of new weapons systems.[30] The French programme instead only emphasizes the simulation of nuclear tests, although there are some rumours concerning a possible opening of the programme to civilian European researchers.[31]

THE NATIONAL IGNITION FACILITY

From the mid-1960s, researchers have studied the use of powerful laser beams to implode to thermonuclear conditions a small target containing a mixture of deuterium and tritium, the heavy isotopes of hydrogen, which is the most reactive fusion fuel.[32] The goal is to achieve thermonuclear ignition and hence the 'explosion' of a mass of fuel of the order of one to a few milligrams. Such Inertial Confinement Fusion (ICF) could be a source of Inertial Fusion Energy (IFE) for future civilian and commercial use. It is now believed that ignition can be achieved by using an ultraviolet laser system with about 200 beams releasing an appropriately shaped pulse of 1.5 MJ in a few nanoseconds. These are indeed the design parameters for the NIF and *Megajoule* lasers.

Since its conception, inertial confinement fusion has also been considered a tool for weapon simulation.[33] This was the main reason why ICF was for a long time classified, at a time when some principles of thermonuclear weapons were still secret. However, this was probably more a matter of establishing a principle of secrecy at a time when nuclear testing was the predominant research tool, and when ICF's parameters were far from those necessary for significant simulations.

Some of the processes occurring in inertial confinement fusion and in an H-bomb are clearly similar. The H-bomb is based on the implosion of a secondary system, which contains the thermonuclear fuel, using the energy released by the explosion of a primary, which is a 'normal' atomic bomb. However, the ICF experiments are different from H-bombs because ICF is triggered by laser beams, and because the thermonuclear fuel is very small and sufficient only to produce a 'micro-explosion' instead of a devastating explosion. Moreover, it should be emphasized that the ignition mechanism of ICF targets is substantially different from that of an H-bomb, where the fissile components (completely absent in ICF) play a fundamental role in the secondary explosion. Finally, the thermonuclear fuel itself is different: ^{2}H and ^{3}H (deuterium and tritium) in ICF, and ^{6}Li^{2}H (lithium deuteride) in the H-bomb. ICF experiments do not allow the study of the fissile primary, a key component of any nuclear weapon.

Following the proposal to build the NIF, weapon simulation by ICF has recently been extensively debated.[34] It is true that certain physical conditions can be only be simulated through microfusion and that for the first time ICF could make possible laboratory experiments on matter under the conditions of temperature and pressure which are achieved in thermonuclear explosions. These results will possibly lead to some adjustments in the computer codes used for weapon simulation.

Is the National Ignition Facility an example of a dual-use system? As pointed out elsewhere in this book (see Chapters 1 and 4), nearly any technology can have dual uses. However, the question is whether, and how, the fact that NIF has been intended for dual use has affected its design and its effectiveness with respect to purely scientific/civilian uses. The requirement of reaching microfusion conditions has determined the megajoule design parameter of NIF. As a consequence, NIF may prove the feasibility of laser-driven fusion ignition, but may not be directly relevant for scaling up towards a laser facility which can produce civilian commercial energy. Such systems will probably need more efficient and cheaper technologies, which are not a part of NIF and are not yet known, and may require simpler ignition schemes.

It could be said that the demonstration of laser-driven fusion ignition will be a brilliant scientific result. However, it will probably come too early to be exploited in practice. Therefore, for the moment a slightly smaller system may be sufficient for scientific research alone. However, the present budgetary restrictions have already forced scientists to design NIF on the basis of innovative technologies, resulting in a cost much lower than it would have been if the design were based on older *Nova* technologies.

Understandably, the National Ignition Facility has been the aspect of the stewardship programme most widely debated by the public, even though it plays a relatively small part in the SBSS programme. A facility such as NIF, which has also many important scientific and civilian applications, may arouse more enthusiasm and attract more support than other more technological elements of the programme. In a way NIF, whose relevance to stewardship is questionable, has played an important role in shifting the attention of the public from nuclear tests to other stewardship issues. Also, NIF is exploiting the competence of the US National Laboratories for projects with potentially high scientific and civilian impacts.

THE RISK OF NEW NUCLEAR PROLIFERATION

The goal of the Comprehensive Test Ban Treaty is to halt nuclear proliferation and the nuclear arms race. The Non-Proliferation Treaty already required the commitment of the nuclear powers to gradual nuclear disarmament.[35] In this framework, the real questions are:

1   Can a nuclear state develop new weapons without testing?
2   Can non-nuclear states develop nuclear weapons without testing?

These two questions are evidently related, respectively, to the 'vertical' proliferation of nuclear arms races between nuclear weapons states, and to the 'horizontal' proliferation of spreading nuclear weapons to additional countries.

Vertical nuclear proliferation

The concern that giant lasers, such as *Megajoule* or NIF, may promote vertical nuclear proliferation has been expressed, for instance, by Sebastian Pease, former director of Culham Research Laboratories in England and a Pugwash member. He said that such lasers only 'serve the purpose of maintaining the technology structure of nuclear laboratories in France and the US. They have the potential for developing weapons and ... if [one] can make explosions without a fission trigger, there could be a development of a new range of fearful things'.[36] The same view is held by Nobel Laureate Georges Charpak of CERN.[37]

But not everybody shares this concern. For instance, the physicist and former president of Pugwash, Josef Rotblat, and many other experts think that such lasers will be much more important for energy research than military use.[38] They feel that it is not possible to fully simulate nuclear explosions in the laboratory and then extrapolate such results to nuclear weapons. They argue that weapons and simulations are based on different physical principles, and that weapons yield energies millions of times larger: for example, a laboratory laser-induced microfusion explosion will release at most 20 Megajoules, equivalent to 5 kg TNT, while the smallest nuclear explosions are instead equivalent to about 1 kiloton, or 4 million Megajoules.

It must be stressed that the high level of development and the diversity reached by the existing nuclear arsenals makes it difficult to think that now, at the end of the Cold War, it is necessary or useful to develop completely new weapons. Mutual Assured Destruction still holds, and the introduction of a new category of weapons cannot change the stable balance between nuclear superpowers. In a way, this conclusion is also implicit in the debate about the Strategic Defense Initiative: there is no technical way out of nuclear deterrence,[39] even if political ways out are possible.

Horizontal nuclear proliferation

In horizontal proliferation, the key element is always the availability of a fissile primary system. This primary cannot be simulated on a reduced scale, and requires the development of expensive and complex technologies for the production of plutonium and enriched uranium. As far as the psychological aspects of the impact on international public opinion are concerned, a newly proliferating country would probably gain the same 'benefits' from the development of fissile bombs as from progressing on to thermonuclear weapons. In other words, no country would feel safer because it is threatened by a country which 'only' has atomic fission weapons instead of thermonuclear weapons.

The acquisition of atomic fission weapons is sufficient to give a country the status of a nuclear power, yet it is easier than acquiring thermonuclear weapons, and is in any case essential for developing them in the future. No matter what, it would not seem a wise idea to try to develop H-bombs starting with laser experiments; indeed, while many thousand thermonuclear bombs have already been produced, laboratory laser ignition has yet to be demonstrated. The case of India and Pakistan proves that if a country is really determined to have atomic weapons, then it will focus its research programmes on topics other than laser research.[40]

It is true that high-energy experiments with lasers could, for instance, give useful information on such topics as the equation of state for plutonium. The diffusion of such scientific knowledge is unavoidable, and will be even more unavoidable in the future. In such scientific matters, it would seem more advisable to try to control the materials, such as fissile materials and tritium, and the technologies, rather than the knowledge itself. Opening major laser laboratories, not only the French and US ones, for non-secret public scientific research would in the end give the best and final assurance against their use for possible nuclear proliferation.

THE FUTURE OF NIF AND *MEGAJOULE*

Owing to its links with thermonuclear weapons, research on inertial confinement fusion (ICF) and inertial fusion energy (IFE) has long suffered from secrecy in the USA and the other nuclear powers. The more recent attitude of the US Department of Energy is quite different: ICF results have been almost completely declassified, which seems to prove the low relevance of ICF research to nuclear proliferation.[41] But how can ICF be of low relevance to proliferation and at the same time be part of stockpile stewardship?

In particular, the only part of the SBSS programme connected to microfusion is the construction of the NIF laser. This is surely the most exciting part of the stewardship programme from the scientific point of view, but it is also of questionable relevance to the broader SBSS objectives. Ray Kidder, a former prominent researcher at Lawrence Livermore National Laboratory, has recently said that 'as far as maintaining the stockpile is concerned, NIF is not necessary'.[42]

If this is true, then will the realization of giant lasers only have the positive effect of stimulating the laser industry and ICF/IFE-related research? There are really two different attitudes among the scientists who do not believe in the relevance of the giant lasers to the stockpile stewardship programme. The first attitude assumes that it will be possible to use military funding for the construction of giant lasers and then use these for civilian research. In contrast, the second attitude emphasizes how the links with military research have so far negatively affected IFE research.

Without the justification of nuclear explosion simulation, would the NIF or *Megajoule* project receive the necessary funding? The cause of energy production through controlled nuclear fusion deserves public support, but even today, both politicians and the public appear to be keener to spend money on defence programmes than on basic scientific research. The political success of the NIF project, and of the SBSS programme in general, is due to a convergence between US conservatives and Democrats which made it possible to start a billion-dollar programme during a period of budgetary restriction. Democrat supporters of the Comprehensive Test Ban Treaty do not want to create doubts about the post-treaty US nuclear capability. Despite having sometimes presented the treaty as a threat to national interests, the conservatives cannot fail to support the directors of the nuclear national laboratories when they say that they need the National Ignition Facility to defend such national interests.

The US Energy Secretary, Ernesto Peña, declared that 'if you support the Comprehensive Test Ban Treaty then you should support NIF'.[43] Many US scientists share his opinion, including such prominent experts as Hans Bethe, Henry Kendall and Herbert York.[44]

Understanding the French *Megajoule* projects requires a slightly different perspective. The openness of the programme, and hence the expected scientific results, promises to be much smaller than that of the NIF project, and the costs are expected to be higher. Are two installations of this kind needed, or would it be more effective to have a single large installation where scientific research would be conducted in an open atmosphere through international collaboration, combined with some additional smaller systems (for example, with laser power of a few hundred kilojoules each) for scientific research on complementary topics? However, due to the present organization of this branch of research, these are questions which are probably going to remain unresolved.

THE NUCLEAR SAFETY AND RELIABILITY PROBLEM

Although this chapter has already addressed issues of reliability and safety, and although such concepts seem intuitively clear, a more precise definition of them may be useful when looking at these issues in detail. For instance, the US Department of Energy (DOE) definition relates 'safety' to accidents and health and pollution hazards, and 'reliability' to the operation of nuclear bombs as weapons.

Recently the DOE has analysed the problems that have occurred in the US nuclear stockpiles and has found that since 1945 there have been 2,400 problems, which were classified into 252 different types.[45] Sixty-six of these types of problems were related to safety, 58 were due to faults in the weapon design, and 8 were connected to ageing. Six of the latter category of faults were found in old warheads which have now been retired from the stockpiles, and none was due to the nuclear components.

The problems related to reliability were of 186 different types, and 164 of these were serious enough that corrective actions had to be taken. Here too, only a small proportion of the problems concerned ageing (24 per cent) and the nuclear components of the weapons (19 per cent). When we analyse such problems quantitatively, the following situation appears. The majority of problem types, 112, reduced reliability by less than 1 per cent; 37 types reduced reliability by 1–5 per cent, six types by 5–10 per cent, and nine types by more than 10 per cent. Here, the reduction in weapon reliability is defined as a reduction of the probability of a successful performance of the weapon at the moment of the explosion. 'Unreliability' here means that the weapon will produce a significantly reduced yield, not that it will fail altogether. Defined in such a way, weapon reliability is more connected to the notion of fighting a nuclear war than to deterrence. For deterrence, the yield of nuclear weapons is not as critical as it would be in the case of a first strike against hardened silos containing ICBMs.

Most of the reliability problems concerned old weapon models. This is mainly due to the fact that during the 'hot' period of the Cold War, rapid mass production of nuclear weapons was undertaken, with the goal of producing as many warheads as possible, without much attention to safety and reliability.

In any case, only a small fraction of the observed problems were related to the nuclear parts and/or to weapon ageing. This strongly suggests that the 'stewardship' programmes should focus on studying and checking the non-nuclear parts of the weapons, such as the 'normal' chemical explosives, electronic and triggering systems, and the delivery systems, as described by Richard Garwin.[46]

At first sight then, the science-based stockpile stewardship, as it is presented now with much emphasis on hydrodynamics, high-energy density and microfusion experiments, seems really inappropriate for its intended purpose. Opponents of the SBSS argue that it should be replaced by a smaller, less costly programme that concentrates on non-nuclear and non-ageing problems. This is a technical matter concerning the contents of the SBSS, rather than whether these should be a SBSS.

There are at least three points about the stewardship programme which are very important:

1   The aspects of the SBSS which are most widely discussed (such as those concerning the NIF project) are not those which consume most of the funding. This is probably due, as discussed earlier, to the need to obtain strong public support for the whole stewardship programme by emphasising scientific research and civilian 'spin-offs'.
2   The DOE assessment of safety and reliability considers a time span which may be long compared to the duration of the Cold War, but is probably shorter than the period during which nuclear weapons must continue to be safeguarded. It is likely that different and new problems related to long-term weapon ageing will appear during that longer period.
3   The SBSS programme is not only scientific and technological, but also plays an important political role in supporting the acceptance of the Comprehensive Test Ban Treaty. In this respect, a smaller programme might not have the same positive impact.

It is worth pointing out once more that the cost of the SBSS programme is not really so high when compared to either the nuclear weapons programmes or the R&D investments made in many strategic industrial fields. It may be a reasonable cost when compared to the resources Western nations have spent on past arms races. Of course, the SSBS resources could be used for other purposes, but so could the much larger total weapon expenditures.

CONCLUSION

The central goal of any well-calibrated stockpile stewardship programme should be to assure the safety and reliability of nuclear stockpiles. This chapter has presented the basic arguments supporting the feasibility of such a programme that involves periodical checks of the weapons without any nuclear explosion of fissile materials.

The material presented here suggests that stockpile stewardship is necessary, but it is unlikely that it could be used to develop new nuclear weapons based merely on laboratory experiments and tests of separate non-nuclear components. New nuclear weapons need to be tested, and deploying untested weapons would be contrary to established military procedures. This applies not only to new generations of weapons, but also to variations of devices that already exist.

The supporters of science-based stockpile stewardship argue that:

1   it has helped to remove objections against the Comprehensive Test Ban Treaty and nuclear disarmament
2   it does not allow the development of new thermonuclear weapons without nuclear explosions
3   it helps in assuring the safety and reliability of the stockpiles.

The opponents of the SBSS programme point out that:

1   the stewardship allows the national nuclear laboratories to maintain a nuclear-weapon design capability
2   it keeps the emphasis on the importance of nuclear weapons
3   its cost and size are excessive.

In evaluating the debate on the stockpile stewardship, however, one should consider that the SBSS or similar programmes are just one (relatively small) component of the disarmament policy of the nuclear powers. It is true that with the SSBS a certain weapon design capability is maintained, but the future of the nuclear stockpiles is not really related to this capability. If this is the beginning of an era of treaties, disarmament and international collaboration, then such a capability will become less and less important. There should be few illusions: it took only a few years for the USA to develop nuclear weapons, using the relatively primitive technology which was available in the 1940s and at a time of war. Nowadays, the scientific knowledge about nuclear weapons exists and cannot be destroyed, and it would now certainly be quicker and easier to build new nuclear weapons.

The discussion about stockpile stewardship has raised questions about giant lasers and inertial confinement fusion. In the framework of the current simulation programmes, the development of the big lasers such as NIF and *Megajoule* is the most interesting aspect from the scientific point of view, both in the perspective of reaching energy production through nuclear fusion and in view of spin-offs to laser technology and to the laser industry.

It is unfortunate that the large scientific projects such as NIF and *Megajoule* are undertaken with military goals and under military control. The possibility of conducting IFE research in the 'spare time' of these devices does not seem so exciting: a civilian programme explicitly aimed at this goal would certainly be more effective. Finally, it is worth observing that although the cost of the NIF or *Megajoule* projects is high, it is smaller than that of some other large scientific experiments or technological projects, such as some particle accelerators, the ITER magnetic fusion reactor and the space research programmes.

APPENDIX 1: MEGALASERS FOR MICROFUSION

The NIF and *Megajoule* projects will be based on neodymium lasers, and hence will use a rather mature technology that has been successfully used in large operational laser systems. In order to increase their efficiency, the NIF and *Megajoule* lasers will use so-called 'multiple-passage' techniques. The laser pulse produced by a single oscillator will be amplified in a series of laser chains. Flash lamps will pump energy into neodymium-doped glass rods emitting infrared light at 1.053 microns. This light beam will be amplified in 18 amplifiers, each of them made with a 40 cm × 80 cm phosphate glass slab illuminated by powerful light flashes.

Two KDP crystals placed in sequence at the end of the amplifier chain will convert 70 per cent of the infrared light into ultraviolet light at 0.35 microns. This wavelength allows more efficient compression of thermonuclear targets.

The lasers will consist of about 200 beams tubes, each 20 m long, and each producing pulses with a total energy of 2 MJ in a few nanoseconds. The beams will be focused on a small thermonuclear target containing a hollow sphere of deuterium and tritium of about 1 mm radius, and located in the centre of a reaction chamber. The thermonuclear burn of such a target is expected to release about 20 MJ.

Currently, the two biggest laser systems are in the USA and are both based on neodymium glass technology. *Omega* has been operating since 1995 at the University of Rochester. It has 60 beams which each produce 40 kJ of ultraviolet light in 1–2 nanoseconds. It is mainly used for direct-drive laser fusion experiments. The ten-beam *Nova* has been operating at the Lawrence Livermore National Laboratory since 1985, producing pulses of 30 kJ of green light.[47]

In France, currently the most powerful laser is *Phebus*, at the CEA laboratory in Limeil, near Paris. Based on *Nova* technologies, it has only two beams, delivering about 7 kJ of their total energy within 1 nanosecond or so. Table 6.1 summarizes the differences between *Phebus* and the future *Megajoule* system, and thereby gives a clearer idea of the scaling up required for the lasers of the SSBS programmes. France also has a medium-size laser system for civilian use, at the LULI Laboratory, at the Ecole Polytechnique near Paris, with a final beam energy slightly less than one kilojoule.

Table 6.1   Comparison of the *Phebus* and *Megajoule* giant lasers

	Phebus	*Megajoule*
Number of component beams	11	110
Weight of Nd-doped glass (tons)	1.2	125
Weight of normal glass (tons)	5.6	66
Weight of KDP crystals (tons)	0.08	2.3
Finished optical surfaces (m^2)	40	9,600
Number of flash lamps	1,200	12,480
Stored electrical energy (MJ)	14	440

Source: P. Hill, 'Nobel prize winners hope for check on large lasers', *Opto and Laser Europe*, No. 41 (June 1997), p.17.

In Japan, the Laboratory for Laser Energetics of Osaka University hosts the *Gekko* XII laser, with 12 beams and a total light energy per pulse of 12 kJ. It is only used for civilian research.[48] The UK has two laser systems with energies of about 1 kilojoule. The first one at the Rutherford Appleton Laboratory, near Oxford, is for civilian applications, the other is for military applications and is located at the Atomic Weapons Establishment in Aldermaston. In Russia, the biggest system is ISKRA 5 at the Arzamas-16 nuclear research centre. A chemical laser, with iodine as the active medium, it produces several kilojoules of beam energy per pulse in the 1.2 micron infrared region.[49]

APPENDIX 2: INERTIAL CONFINEMENT FUSION

In nuclear fusion reactions, two light nuclei merge ('fuse') to form a heavier nucleus, releasing a large amount of energy. Such reactions are the energy sources of the stars and thermonuclear weapons (H-bombs), and will possibly be used in future controlled thermonuclear fusion reactors for energy production. In the basic fusion reaction in stars, two protons (two hydrogen nuclei) merge to form deuterium (the nucleus of a heavier isotope of hydrogen). In research on controlled thermonuclear fusion,[50] a fuel mixture of deuterium (D = ^{2}H) and tritium (T = ^{3}H, another heavy isotope of hydrogen) is used. The fusion reaction is orders of magnitude more rapid than that for ordinary hydrogen, and the reaction results in the production of a helium nucleus, a neutron and large amounts of energy:

$$^2D + {}^3T \rightarrow {}^4He + {}^1n + 17.6 \text{ MeV of energy}$$

In order to fuse, the two nuclei must collide. This requires that they possess a kinetic energy sufficient to partially overcome their electric repulsion. In practice, to achieve a rapid fusion reaction the D + T fuel must be heated to hundreds of millions degrees centigrade – which explains the term 'thermonuclear' fusion.

To achieve an effective fusion reaction of significant amounts of fusion fuel, another condition has to be satisfied concerning 'confinement' of the burning fuel. This confinement can be achieved either by containing the fuel in a properly shaped magnetic 'bottle' or 'doughnut', or by compressing the fuel to very high density. Such compression is the fundamental element of Inertial Confinement Fusion, where the D + T mixture is compressed to a density about 1,000 times that of liquid hydrogen, corresponding to pressures of a dozen billion atmospheres. No external means can be used to confine the reacting material, which stays assembled for a brief time only, due to its own inertia.

In direct-drive laboratory experiments, a hollow spherical target containing a layer of D + T is irradiated by powerful laser beams. The external layers are rapidly evaporated, and consequently the remaining target material is accelerated and implodes. When the empty space in the middle of the target is filled by the fusion fuel under compression, the conditions of pressure and temperature needed for ignition are reached. A target may contain a few milligrams of D + T, and will yield an energy of some hundreds of Megajoules. The fusion of the target material will be explosive, and will last about one-tenth of a nanosecond.

The main problem in current ICF experiments is the uniformity of the compression. If the laser beams do not irradiate the target uniformly, or if hydrodynamic instabilities amplify small target defects too much, it is then impossible to achieve ignition. As a possible solution to this problem, many laboratories, including the Lawrence Livermore National Laboratory, are studying a technique called 'indirect-drive' inertial confinement fusion. In this technique, the laser beams are not directly focused onto the fusion capsule, but on the walls of a bigger cavity which encloses the capsule itself. The material of the wall is heated to high temperature and emits low-energy X-rays. Such radiation is confined in the cavity acting as an oven, and produces a uniform compression of the capsule. This solution relaxes the homogeneity requirements for the laser beams, but requires greater laser energy.

ICF physics has been studied since the mid-1960s. As indicated above, experiments, theories and simulations support the prediction that an ultraviolet laser system delivering an energy of 1.5 MJ, with peak power of about 500 TW and proper pulse shape, should be adequate for achieving ignition. As in every scientific experiment, there are uncertainties in this prediction. In order to reduce any possible negative surprise, in 1986 and 1987 the USA carried out a series of experiments to study the ignition of a thermonuclear capsule whose implosion was driven by the X-ray radiation released by underground nuclear explosions. These 'Centurion-Halite' tests were conducted under the classification regime, but a report of the US National Academy of Science, which had access to the results, stated that such results 'qualitatively showed that the physical concepts at the basic of ICF are sound'.[51]

Despite all the differences between inertial confinement fusion and H-bombs, ICF research conducted by the nuclear powers has been classified because of the physical similarities between the implosion of the ICF capsule and that of the secondary of an H-bomb. This has strongly affected ICF research: for instance, some thirty years ago the European Community Agency Euratom chose not to support any research in this field because of its military implications. In practice, the situation has never been completely clarified: direct-drive experiments have always been subjected to fewer restrictions, basic research on radiation confinement was conducted at a few open laboratories in Germany and Japan, and Japanese groups even performed indirect-drive implosion experiments.

Recently, the US Department of Energy has revised its policy on the relevance of inertial confinement fusion to nuclear proliferation.[52] Consequently, most ICF results have been declassified and published in many detailed papers, including a special issue of the scientific journal *Physics of Plasmas*.[53] Nowadays, classification is restricted only to some details of computer codes and some aspects of equations of state and radioactive properties of matter at high temperature and density. In the mean time, ICF laboratories have been partially opened to the external scientific community.

ACKNOWLEDGEMENT

The authors are indebted to Paolo Farinella and Dietrich Schroeer for very useful discussions on science-based stockpile stewardship.

REFERENCES

1   S. Glasstone and P.J. Dolan (eds), *The Effects of Nuclear Weapons* (Washington, DC: US Government Printing Office, 3rd edn, 1977). D. Schroeer, 'The fusion bomb', in *Science, Technology and Nuclear Arms Race* (New York: Wiley, 1984), pp.58–81.
2   See the contributions on proliferation in this volume by Dingli Shen (Chapter 7) and Patricia Lewis (Chapter 8) and the classic article by W. Epstein, 'The proliferation of nuclear weapons', *Scientific American*, Vol. 232, No. 4 (April 1975), pp.18–33.
3   W. Sweet, 'Faults and failures: The Indian and Pakistani tests – verification breakdown or startling success?', *IEEE Spectrum*, Vol. 35, No. 7 (July 1998), pp.87–8.
4   R. Garwin, 'Les essais nucleaires ne sont plus necessaires' ('Nuclear tests are no longer necessary'), *La Recherche*, Vol. 282 (December 1995), p.70.
5   R. Galy-Dejean, 'La simulation des essais nucleaires' ('The simulation of nuclear tests'), *Rapport d'information* (French National Assembly), No. 847 (December 1993).
6   Garwin, 'Les essais nucleaires?'.
7   H.F. York, 'The great test-ban debate', *Scientific American*, Vol. 227, No. 5 (November 1972), pp.15–23. H.R. Myers, 'Extending the nuclear-test ban', *Scientific American*, Vol. 226, No. 1 (January 1972), pp.13–23.
8   Garwin, 'Les essais nucleaires', *op. cit.*
9   'O'Leary gives the green light to National Ignition Facility', *Fusion Power Report*, Vol. 15, No. 10 (October 1994), p.97.
10 S. Drell, et al., 'Science Based Stockpile Stewardship', *Mitre Corporation Jason Report JSR-94-345* (November 1994).
11 Galy-Dejean, 'La simulation'.
12 'House Majority leader tours LLNL, offers guarded support to NIF', *Fusion Power Report* (August 1995), p.77. 'DOE report gives thumbs up to NIF arm control goals', *Fusion Power Report* (August 1995), p.73.
13 H. Zerriffi and A. Makhijani, 'The stewardship smokescreen', *Bulletin of the Atomic Scientists*, Vol. 52, No. 5 (September/October 1996), p.23.
14 J. Murray et al., 'Special issue: National Ignition Facility design and activities', *ICF Quarterly Report*, Vol. 7, No. 3 (April/June 1997).
15 Commissariat à l'Energie Atomique (CEA), *Annual Report 1997*, (Paris: CEA Direction de la Communication, 1997); see the section 'Science at the service of defence' pp.24–26. See also CEA, 'Vers la combustion du DT par laser' ('Towards the Ignition of DT by Laser') in *CHOCS revue Scientifique et technique de la direction des Applicationes Militaries*, No. 13 (Paris: CEA, April 1995), pp.5–84.
16 'Essais nucleaires: Giscard prend position ('Nuclear tests: Giscard takes up position'), *Le Figaro*, 12 October 1995.
17 'L'heure de la simulation ('The time for simulation'), *Le Figaro*, 31 January 1996.
18 'La France se dote d'un laser géant pour simuler les essais nucleaires' ('France acquires a giant laser to simulate nuclear tests'), *Le Monde*, 5 May 1995.
19 Drell et al., 'Science Based Stockpile Stewardship'. S. Atzeni, 'Esperimenti di

laboratorio, microfusione e CTBT' ('Laboratory experiments microfusion and CTBT'), in F. Paoan (ed.), *L'impegno italiano per il controllo internazionale degli armamenti nucleari* ('The Italian Commitment to the International Control of Nuclear Weapons') (Roma: Ente per le Vuove Tecnologie l'Energia e l'Ambiente, 1996), pp.243–66.

20   Zerriffi and Makhijani, 'The stewardship smokescreen'.
21   Ibid.
22   CEA, *Annual Report 1997*. CEA, 'Vers la combustion'. Atzeni, 'Esperimenti di laboratorio'.
23   Zerriffi and Makhijani, 'The stewardship smokescreen'.
24   F. von Hippel and S. Jones, 'Take a hard look at subcritical tests', *Bulletin of the Atomic Scientists*, Vol. 52, No. 6 (November/December 1996), p.44.
25   Atzeni, 'Esperimenti di laboratorio'.
26   R.C. Von Hippel and Jones, 'Take a hard look'.
27   See, for instance, D. Batani, M. Koenig, T. Lower, A. Benuzzi and S. Bossi, 'Measuring equation of state with lasers', *Europhysics News*, Vol. 27 (1996), p.210, and references therein.
28   Drell et al., 'Science Based Stockpile Stewardship'. Atzeni, 'Esperimenti di laboratorio'. See also J. Glanz, 'Will NIF put the squeeze on Sandia's Z Pinch?', *Science*, Vol. 277 (18 July 1997), p.306.
29   Atzeni, 'Esperimenti di laboratorio'.
30   'O'Leary gives the green light'. 'DOE report gives thumbs up to NIF arm control goals', *Fusion Power Report* (August 1995), p.73.
31   D. Batani and S. Atzeni, 'Megalaser e microesplosioni: si può davvero simulare una bomba H in laboratorio?' ('Megalaser and micro-explosion: Is it possible to simulate H-bombs in the laboratory?'), *Sapere*, No. 6 (December 1997), p.38.
32   J.W. Nuckolls, L. Wood, A. Thiess, and G.B. Zimmerman, 'Laser compression of matter to super-high densities: Thermonuclear (CTB) applications', *Nature*, Vol. 239 (1972), p.139. See also J.L. Emmett, J.W. Nuckolls and L. Wood, 'Fusion power by laser implosion', *Scientific American*, Vol. 230, No. 6 (June 1974), pp.24–37.
33   M. Tobin, L. Choate and D. Beller, 'Use of inertial confinement fusion for weapons effects simulations', *ICF Quarterly Report*, Vol. 2, No. 4 (July/September 1992), p.194.
34   'Weapons-relevant NIF research areas center on basic physics', *Fusion Power Report* (August 1995), p.75. See also S.B. Libby, 'NIF and national security', *Energy and Technology Review* (December 1994), p.23; G.L. Lubkin, 'NIF and stockpile stewardship', *Physics Today*, Vol. 48, No. 8 (August 1995), p.24; J. Glanz, 'A harsh light falls on NIF', *Science*, Vol. 277 (18 July 1997), p.304.
35   A. Myrdal, 'The international control of disarmament', *Scientific American*, Vol. 231, No. 4 (October 1974), pp.21–32.
36   P. Hill, 'Nobel prize winners hope for check on large lasers', *Opto and Laser Europe*, No. 41 (June 1997), p.17.
37   Ibid.
38   Glanz, 'A harsh light'. D.H. Crandall, 'NIF: harsh light or illumination?', *Science*, Vol. 279 (22 July 1997), p.1,021. See also, for example, A.B. Carter, *Directed Energy Missile Defense in Space*, background paper (Washington, DC: Office of Technology Assessment, April 1984); R.L. Garwin, K. Gottfried and H.W. Kendall, *The Fallacy of Star Wars* (New York: Vintage Books, 1984).
39   D. Batani, F. Lenci and G. Colombetti, 'Missili balistici e sistemi di difesa antimissilistici' ('Ballistic missiles and anti-missile defence systems'), in *Cinquant'anni dopo Hiroshima* ('Fifty Years After Hiroshima') (Rome: CESPI-USPID, 1995).

40 D. Sharma (ed.), *The Indian atom: Power and proliferation* (New Delhi: Philosophy & Social Action, 1986); see also note 4 above.

41 'DoE declassifies fusion research', *Laser Focus World*, Vol. 30, No. 2 (1994), pp.22–3.

42 Glanz, 'A harsh light'.

43 I. Goodwin, 'Peña breaks ground at Livermore laser-fusion facility ...,' *Physics Today*, Vol. 50, No. 8 (August 1997), p.46.

44 Glanz, 'A harsh light'. Hill, 'Nobel prize winners hope'. Goodwin, 'Peña breaks ground'.

45 Zerriffi and Makhijani, 'The stewardship smokescreen'.

46 Garwin, 'Les essais nucleaires'.

47 Batani and Atzeni, 'Megalaser e microesplosioni'. Hill, 'Nobel prize winners hope'.

48 Batani and Atzeni, 'Megalaser e microesplosioni'. Hill, 'Nobel prize winners hope'.

49 Batani and Atzeni, 'Megalaser e microesplosioni'. Hill, 'Nobel prize winners hope'.

50 See, for instance, J. Meyer-ter-Vehn, S. Atzeni and R. Ramis, 'Inertial confinement fusion', *Europhysics News*, Vol. 29 (1998), pp.202–5; G. Velarde, Y. Ronen and J.M. Martinez-Val (eds), *Nuclear Fusion by Inertial Confinement: A Comprehensive Treatise* (Boca Raton, FL: CRC Press, 1993); W.J. Hogan (ed.), *Energy From Inertial Fusion* (Vienna: International Atomic Energy Agency, 1995).

51 Glanz, 'A harsh light'.

52 'DoE declassifies fusion research'.

53 J.D. Lindl, 'Development of the indirect drive approach to inertial confinement fusion and the target physics basis for ignition and gain', *Physics of Plasmas*, Vol. 2 (1995), pp.3,933–4,024.

PART II
MILITARY TECHNOLOGY
TRANSFER TO
LESS-DEVELOPED COUNTRIES

7   Promoting Non-proliferation: A Chinese View

Dingli Shen

INTRODUCTION

This chapter addresses recent Chinese attitudes toward nuclear non-proliferation. In particular, it discusses China's policies and efforts in nuclear export control, its stance on a possible fissile cut-off treaty, its consistent call for complete prohibition and thorough destruction of nuclear weapons, and its unconditional provision of both positive and negative security assurances.

Proliferation of weapons of mass destruction and their means of delivery has been regarded as a major threat to international peace and security. In this regard, the international community has made great efforts to combat proliferation by permanently extending the nuclear Non-Proliferation Treaty (NPT), negotiating and concluding the Comprehensive Test Ban Treaty (CTBT), and making the Chemical Weapons Convention (CWC) Treaty enter into force.

China has contributed significantly to strengthening international non-proliferation regimes, supporting the indefinite extension of the NPT, working with other parties to achieve the CTBT, and being the second nation to sign it. China has also signed and ratified the CWC, helping to strengthen its worldwide authority.

Notably, China has greatly strengthened its nuclear export control, publishing its regulations on such control and joining the Zangger Committee, and has also made it clear that it will promote a ban on the production of fissile materials for nuclear weapons or other nuclear explosive devices. China has called for a deeper nuclear disarmament leading to the complete prohibition and thorough destruction of all nuclear weapons, since it considers that nuclear disarmament would encourage nuclear non-proliferation. In particular, China deems it important to provide both positive and negative security assurances, unconditionally. This chapter will address these issues.

NUCLEAR EXPORT CONTROL

In general, China has three objectives in terms of nuclear issues:

1   advancing nuclear disarmament
2   preventing nuclear proliferation
3   promoting the peaceful use of nuclear energy.

Due to the dual-use nature of nuclear materials, China has adopted the following three principles regarding nuclear exports. Whatever China exports:

1   shall be used for peaceful purposes only
2   should be subject to safeguards from the International Atomic Energy Agency (IAEA)
3   should not be transferred to a third country without China's prior approval.

In the past few years, China has made a series of commitments to strengthening nuclear export control. On 11 May 1996, China's Ministry of Foreign Affairs indicated that China would refrain from providing assistance to nuclear facilities beyond the safeguards of the IAEA, clearly attaching importance to safeguards. In May 1997, China's State Council promulgated the 'Circular on Strict Implementation of China's Nuclear Export Policy', providing that nuclear materials, nuclear equipment and related technology, as well as non-nuclear materials for reactors and nuclear-related dual-use equipment, materials and relevant technology on China's export control list must not be supplied to nor used in nuclear facilities not subject to IAEA safeguards.

On 1 August 1997, the Chinese State Council reviewed and approved, in principle, Decree No. 230, 'PRC Regulations on the Control of Nuclear Exports', and the attached control list of nuclear exports, identical to the Zangger Committee's trigger list (Part I of INFCIRC 254). On 10 September, the State Council published Decree No. 230. Five days later, the Ministry of Foreign Affairs announced that China had decided to join the Zangger Committee, and on 16 October, China officially joined the committee. On that occasion, China made a commitment that by mid-1998 it would complete its regulations on export control of nuclear-related dual-use items, with the control list to be the same as Part II of INFCIRC 254 on dual-use items. On 1 June 1998, the Chinese State Council reviewed and approved, in principle, the 'PRC Regulations on the Export Control of Nuclear-Related Dual-Use Items and Related Technology'.

It is worth mentioning that these regulations have the following provisions:

1   Nuclear exports from China are monopolized by units earmarked by the State Council. No other unit or individual is allowed to engage in such exports. A licensing system is applied to nuclear exports by the state – it is necessary to apply for an export licence for all the items and their technologies on the 'Nuclear Export Control List'.

2   The China Atomic Energy Authority (CAEA) shall examine nuclear export applications. If the approved application involves nuclear materials, it shall be referred to the Commission of Science, Technology and Industry for National Defence (COSTIND) for re-examination; if it involves other materials, it shall be referred to the Ministry of Foreign Trade and Economic Co-operation (MOFTEC) for re-examination. For any nuclear export which has a bearing on state security, social and public interests or foreign policy, the Ministry of Foreign Affairs should be consulted; when necessary, such exports should be further reported to the State Council for approval. After approval, MOFTEC will issue an export licence.

3   If the recipient government violates its commitment made under these regulations, or if there is an imminent danger of nuclear proliferation, the relevant department of the Chinese Government has the right to suspend such exports.

4   Any violation of the regulations is liable to punishment by law.[1]

The whole process of China's strengthening of nuclear export control apparently had an impact on Chinese–US relations as well. On 29 October 1997, the two governments issued a joint statement during the Jiang–Clinton summit in Washington, DC, indicating that the two countries have respectively taken necessary steps to implement the 1985 Chinese–US Agreement for Nuclear Co-operation. In the words of President Clinton, this is a 'win–win' result, as it assists US efforts to promote non-proliferation, increases opportunities for US nuclear business in China, and enhances environmental protection. On the part of China, it made the certification of the 1985 nuclear agreement a major focus for the summit meeting, and consequently built a constructive mood for the success of the summit and state visit to the USA.

On 12 January 1998, President Clinton signed the formal certification and reports required by US law to implement the Chinese-US Agreement for Peaceful Nuclear Co-operation. He declared that China had met the 'nuclear nonproliferation requirements and conditions necessary under US law to engage in peaceful nuclear cooperation with US industry'.[2] The approval process in the US Congress went smoothly, and the agreement became effective on 19 March 1998.

Symbolically, the certification of the agreement demonstrates that the US policy of engaging China is achieving concrete results in seeking co-operation from China on arms control and non-proliferation. While allowing China to secure for civilian use what is probably one of the best sources of nuclear energy (in terms of advanced technology – especially high safety standards), it also provides the US nuclear industry access to the lucrative Chinese market of nuclear energy development. Nevertheless, one has to be cautious as to how great a share US nuclear companies can win of the Chinese market – probably several billion US dollars out of the dozen-billion-dollar market.

THE FISSILE CUT-OFF TREATY

After the CTBT, a Fissile Material Ban (FissBan) is commonly expected as the next step in nuclear arms control and non-proliferation. On 4 October 1994, China issued a joint statement with the USA, stating that the two countries: 'in support of their shared interest in preventing the proliferation of nuclear weapons, have agreed to work together to promote the earliest possible achievement of a multilateral, non-discriminatory, and effectively verifiable convention banning the production of fissile materials for nuclear weapons or other nuclear explosive devices'.[3]

By signing the joint statement, it is recognized that a fissile cut-off treaty provides a vehicle for working to halt the production of fissile materials for nuclear weapons or other nuclear explosive devices in key threshold states. At the 49th United Nations General Assembly, China also expressed its support for negotiations on a fissile cut-off treaty.

However, the negotiation of a fissile cut-off treaty has still not commenced at the Conference on Disarmament in Geneva, for reasons including other agenda priorities, and the linkage of a fissile material cut-off to time-bound nuclear disarmament. In spite of this, China consistently supports this FissBan effort, and has formulated some of its policies on this issue:

1   The treaty shall have universality. All nuclear-capable states should join it.
2   The scope of the treaty shall be defined by the purpose for which these fissile materials are produced. Such purposes include military weapons use, other military use (for instance, use in providing nuclear power) and civilian use. The FissBan shall only restrict production of fissile materials for nuclear weapons use.

3   The negotiation of the treaty shall be carried out on a fair, reasonable basis. The rights of signatories, particularly those of developing countries, to use nuclear energy for peaceful purpose and hence develop their economy, shall be fully respected.
4   The treaty's verification should be in accordance with the scope of the treaty, and should be reasonable, effective and practical. Intrusiveness shall be as low as possible in order to avoid abuse of the treaty. Cost-effectiveness shall also be taken into consideration, so that the signatories can bear the financial responsibility in regard to verification.

It is hoped that a new momentum in negotiating the fissile cut-off treaty can be built up in the near future.

NUCLEAR DISARMAMENT

Ever since China developed nuclear weapons, it has called for the complete prohibition and thorough destruction of nuclear weapons worldwide. China holds that countries with the largest and most effective nuclear and conventional arsenals have a special responsibility in arms control and disarmament, and that nuclear disarmament is the responsibility of nuclear weapons states, corresponding to the obligation of non-nuclear weapons states not to acquire nuclear weapons. China considers that the two nuclear weapons superpowers should implement the bilateral nuclear disarmament agreements that they have reached, and commit to further drastic reductions in their nuclear armaments.

At the Helsinki Summit in March 1997, the US and Russian governments addressed an accelerated process to ratify the second Strategic Arms Reduction Treaty (START II), and they even discussed the possibility of START III.[4] Questions have been raised as to whether China will join the nuclear disarmament process, and even further, whether it will dispose of nuclear materials from dismantled nuclear weapons.

As a nuclear weapons state, China has a responsibility for nuclear disarmament. However, this has to be addressed in a multilateral framework. Although China's nuclear weapons programme has been veiled in secrecy for the purposes of deterrence, it is understood from sources in the West that its arsenal is measured in hundreds of weapons. Against this background, the following aspects have to be considered in addressing the possibility of China's participation in a multilateral nuclear disarmament process:

1   The expected START III level of deployed strategic nuclear warheads is still too high to attract medium-sized nuclear weapons states to join.

2   The stockpiled nuclear warheads of the nuclear superpowers must be addressed as well.
3   The fissile materials from dismantled nuclear weapons of the USA and Russia must be treated to render them unusable for weapons.
4   National missile defence and highly capable theatre missile defence undermine the credibility of nuclear deterrence of medium-sized nuclear weapons states and discourage their interest in participating in multilateral processes.
5   All nuclear weapons states, especially the major nuclear weapons states, must commit to a policy of no first use of nuclear weapons.

SECURITY ASSURANCES

China adopted a no-first-use policy from the date it tested its first nuclear weapon. This unconditional assurance was complemented on 5 April 1995, when China issued the *National Statement on Security Assurances*:[5]

1   China undertakes not to be the first to use nuclear weapons at any time or under any circumstances.
2   China undertakes not to use or threaten to use nuclear weapons against non-nuclear weapons states or nuclear-weapon-free zones at any time or under any circumstances.
3   As a permanent member of the Security Council of the United Nations, China undertakes to take action within the council to ensure that the council takes appropriate measures to provide, in accordance with the United Nations Charter, necessary assistance to any non-nuclear weapons state that comes under attack with nuclear weapons, and impose strict and effective sanctions on the attacking state.

China considers that unconditional positive and negative security assurances are conducive to nuclear non-proliferation, and strongly calls for the early conclusion of an international convention on no first use of nuclear weapons as well as an international legal instrument to insure the non-nuclear weapons states and nuclear-weapon-free zones against the use or threat of use of nuclear weapons.

This call has not been responded to by the largest nuclear weapons state, on the grounds of the necessity of maintaining extended nuclear deterrence. Putting aside the controversial nature of the extended deterrence policy, it was reported that the USA has recently adopted new targeting guidelines for nuclear arms, allowing it to broaden the scope of its targeting against China. This has certainly upset China, and raises concerns over the recently built 'constructive strategic partnership' between the two countries.[6] Washington's move is not helpful in seeking broader co-operation in non-proliferation.

CONCLUSION

At a time when nuclear proliferation has become one of the main concerns of international security, China has greatly strengthened its non-proliferation stance. It has tightened domestic controls over nuclear exports, and joined the Zangger Committee to promote control of dual-use nuclear-related items. It has committed to the negotiations for a ban on production of fissile materials for weapons purposes. China has also called for the total elimination of nuclear weapons and for a no-first-use position before this is fulfilled. All these measures help strengthen international efforts to halt nuclear proliferation.

REFERENCES

1 Statement by Ambassador Li Changhe of the Chinese Permanent Mission in Vienna at the Meeting of the Zangger Committee, 16 October 1997.
2 Statement by the Press Secretary on China Nuclear Certification, 15 January 1998; 'US, China close to nuclear co-operation', *China Daily*, 6 February 1998, p.4.
3 *Joint Statement of the United States and the People's Republic of China on Stopping the Production of Fissile Materials for Nuclear Weapons* (Washington, DC, 4 October 1994).
4 In the US-Russian *Joint Statement on Parameters on Future Reductions in Nuclear Forces*, signed on 21 March 1997 in Helsinki, the two sides agreed that the START III negotiations will include four basic components, including a limit of 2,000–2,500 deployed strategic nuclear warheads for each side by the end of 2007.
5 'National statement on security assurances issued', *China Daily*, 6 April 1995, p.1.
6 'Clinton issues guidelines on nuclear arms strategy', *China Daily*, 8 December 1997, p.12. 'China criticizes US for "nuclear deterrent" policy', *China Daily*, 10 December 1997, p.1.

8 Export Controls and Nuclear Weapons

Patricia Lewis

CURRENT EXPORT CONTROLS AND THE THINKING BEHIND THEM

Controls on nuclear materials and nuclear technology began during the Second World War in order to try to prevent the Third Reich obtaining any advantage in what was thought to be a race to make the first atomic bomb. The focus for denial shifted to other countries as the war drew to an end and after it, and the objects of export controls by the West were primarily the Soviet Union and Eastern Europe.

In the 1940s and 1950s, nuclear export controls were devised and implemented on national bases, until 1968, when the nuclear Non-Proliferation Treaty (NPT) was negotiated. The NPT is a bargain between the nuclear weapon states (of which there are five recognized by the treaty) and non-nuclear weapon states. The bargain is two-way. The non-nuclear weapon states agree not to acquire nuclear weapons, and to have their civilian nuclear activities subject to intrusive inspection by the International Atomic Energy Agency (IAEA). The nuclear weapon states agree (along with other states parties to the treaty) to assist states parties with civilian nuclear technology and to negotiate, in good faith, steps to bring about nuclear disarmament.

As part of what has become known as the 'non-proliferation regime', other export control bodies were established. Their main priority has been to support the articles of the NPT with practical export control measures that can be operated on a common basis between a group of like-minded countries.

The Zangger Committee was established in 1971 to establish guidelines for implementing the export control provisions in the NPT. It now consists of over thirty states, and meets twice a year in Vienna. Over the years, the

Zangger Committee has developed a list of sensitive items to be controlled, known as the 'Trigger List'. The committee requires that exports of items on the Trigger List can only be carried out if they:

1 are not to be used for explosive purposes
2 are subject to full-scope safeguards
3 are not re-exported unless they are then subject to safeguards in the second recipient state.[1]

The Nuclear Suppliers Group (the NSG, also called the London Club) was established in 1974, and is independent of the IAEA. The NSG differs from the Zangger Committee in that the participating states agree to exercise restraint in transferring sensitive items, can veto reprocessing and enrichment technology transfer and re-transfers to third parties, have a list of dual-use technologies subject to control, and seek end-use assurances from the recipient states.[2]

There are other controls on sensitive and dual-use technology, including tougher national controls and wider export control regimes such as the Wassenaar Arrangement and the Missile Technology Control Regime.[3]

WHY MIGHT STATES WANT NUCLEAR WEAPONS?

There are a variety of reasons why states might seek to acquire nuclear weapons, including: national security concerns; fears that neighbours may be acquiring a nuclear weapons capability; prestige and status; pressure from domestic opinion; development of the capability by the domestic scientific community; pressure from the military forces, and an intention to gain financial concessions in exchange for halting a nuclear weapons programme.

The nuclear weapon states

From the 1940s to the 1960s, five states developed nuclear weapons. These states, China, France, the UK, Russia (formerly the Soviet Union) and the USA are now legally recognized as the nuclear weapon states.

Reasons for the acquisition of nuclear weapons vary among them, and are complex. In the case of the USA, they were initially developed due to concern over Nazi capabilities, they were later used against Japan, then had a role in deterrence against the Soviet Union. The Soviet Union developed nuclear weapons in response to a perceived threat from the USA. The UK and France developed them for reasons of status and prestige (even though they were already permanent members of the UN Secretary Council), to counter the overwhelming influence of the USA in military affairs, and to have some independent deterrence against the Soviet Union in Europe. China developed them for reasons of prestige and status and as deterrence against the USA and the Soviet Union.

Now that the Cold War is in the past, the interesting question is: why do the nuclear weapon states feel the need to retain their nuclear weapons and, in the case of the USA and Russia, still in large numbers?

The cases of South Africa, Brazil, Argentina and the Ukraine

Although these states all gave up their nuclear weapons capabilities voluntarily, their reasons for having them, or attempting to have them in the first place, differ.

In the cases of Brazil and Argentina, nuclear weapons were proposed as deterrent forces against each other. Pressure from the military and scientific communities of the day were the driving forces. Only when relationships between them improved as a result of major changes in their governments did the two states agree that no longer would they attempt to acquire a nuclear weapons capability. Both Brazil and Argentina are now parties to the Non-Proliferation Treaty and the Treaty of Tlatelolco.

During its Apartheid era, South Africa developed a number of nuclear weapons as a result of its sense of isolation. The government of the day believed that should their state security ever be overwhelmingly threatened, they would carry out a nuclear test as a warning to the opposing forces and to the USA, in the hope that the latter would then be forced to come to their rescue. As the Apartheid regime came to its end, the nuclear weapons were dismantled and the International Atomic Energy Agency was called in to verify that this was indeed the case.

The situation in the Ukraine was altogether different. Left with Soviet nuclear weapons following the break-up of the Soviet Union, the Ukraine seriously considered keeping the nuclear capability as a deterrence against a potentially threatening Russia. It took enormous international pressure, although little in the way of security assurances, to persuade Ukraine that its interests were best served as a *non*-nuclear weapon state.

Israel

The position of Israel in the Middle East is as a small state surrounded by those which it perceives as hostile to its existence. This has led Israel to a position of nuclear weapons ambiguity. It is widely assumed that Israel has a strong nuclear weapons capability, but it has never declared this. Israel links the abolition of all weapons of mass destruction in the Middle East with a peace settlement. Until the peace settlement, Israel will not give up any nuclear weapons capability it may have.

India and Pakistan

Despite a nuclear test by India in 1974 (which it called a 'peaceful nuclear explosion'), both India and Pakistan held a position of nuclear weapons ambiguity in which it was widely assumed that both had the capability but neither declared themselves. In May 1998, India carried out a number of nuclear tests, and Pakistan followed suit. Their capabilities are no longer in doubt.

India claims that it has developed weapons in response to China's nuclear capability and the conflict between China and India in 1962. Pakistan has acquired nuclear weapons in response to India's capability. Since the nuclear tests in May, India's reasons for developing nuclear weapons have been repeatedly articulated. Reasons of national security, status, prestige, equality, a sense of being cornered by the demands of the Comprehensive Test Ban Treaty, and a desire to trigger meaningful nuclear disarmament have all been put forward by India. Pakistan's reasoning has been simply to match India's actions and to demonstrate its capability as a deterrence to India.

Iraq

Despite the cease-fire agreement at the end of the Gulf War and the strenuous efforts to the UNSCOM UN monitoring effort, there is little doubt that Iraq still seeks to acquire weapons of mass destruction. The reasoning behind Iraq's desire for nuclear weapons (which remains unabated although thwarted by export controls) seems to have little to do with status and prestige, or even deterrence. Rather, it looks more likely that they were or are seen as instruments of aggression and political leverage.

North Korea

Assuming that North Korea was attempting to manufacture nuclear weapons, the reasoning behind the programme is not clear, and it seems to have had little to do with prestige and status. Nuclear weapons may have been seen as a deterrent against South Korea or the USA. They may also have been seen as a potential instrument of aggression against South Korea, the USA or Japan. In addition, the capability was certainly used as a means for obtaining state-of-the-art civilian nuclear technology, other forms of energy contributions and a great deal of attention on the problems facing North Korea.

Other possible potential proliferators

There are a large number of potential nuclear proliferators. Many countries have the technologies required; other countries or terrorist groups could seek to purchase nuclear weapons from one or two countries or from militaries within those countries (although chemical or biological weapons may be more useful and credible as a weapon of terror). It seems to be true that most countries with the required nuclear technical capability have no interest in manufacturing nuclear weapons. However, for some of those countries, the tests by India and Pakistan came as a shock and may lead them to rethink their positions on nuclear weapons and non-proliferation, and perhaps to develop contingency plans in case the nuclear proliferation situation worsens.

HAVE EXPORT CONTROLS HAD AN EFFECT ON THE ABILITY TO ACQUIRE NUCLEAR WEAPONS?

Clearly, export controls have had an effect on the ability of some countries which have built or attempted to build nuclear weapons. In most cases, for most sensitive technology, export controls have made the purchase of such technologies either very difficult or impossible. There have been stark exceptions to this denial, however, and the effects of those leakages should not be underestimated.

Take Iraq as an example. During the 1970s and 1980s, Iraq had in place a very active nuclear acquisition programme. When Iraq could not buy the materials and equipment it wanted on the legal and illegal markets, it either sought to manufacture the parts itself or found a different technological route towards nuclear weapons. For example, Iraq used calutrons for uranium enrichment in place of centrifuge technology, although there were a number of centrifuges that it had been able to acquire. Even accounting for the porosity of the export controls against Iraq, its ability to nearly assemble a working nuclear bomb was a painful lesson for supplier states.

South Africa also developed a nuclear weapons programme despite export controls and a hostile international environment, as did India, Pakistan and perhaps Israel, Argentina, Brazil and North Korea.

The main effect of export controls seems to have been to delay and hinder the nuclear weapons capabilities of would-be proliferators. If a country is determined enough and has the technology base, the financial resources and the fissionable material, then export controls alone will not prevent it obtaining a nuclear weapons capability.

The inability of export controls to prevent determined nuclear proliferation could indicate a number of problems:

1   The trigger lists and dual-use technology lists are not adequate or up to date.
2   The technology for manufacturing a simple fission bomb is so old now (dating from the 1940s) that it is available to all countries that want it, regardless of attempts to control it.
3   Export controls are not adequately implemented, as governments of supplier states turn blind eyes to their industries' exports when jobs are at stake, or when they have an understanding (political or economic) with the receiving state.
4   It may be impossible to implement export controls to such a level that they can prevent proliferation.
5   Export controls may only be seen as delaying tactics.
6   Export controls may actually contribute to proliferation by highlighting which technology is important and increasing the allure of nuclear weapons as 'forbidden fruit'.

THE EFFECT OF EXPORT CONTROLS ON THE DESIRE TO ACQUIRE NUCLEAR WEAPONS

Export controls deal only with the supply side of the nuclear proliferation equation. They are available in their application and effect, but can help reduce the flow of sensitive technology to potential proliferators. They are clearly seen as an attempt by the supplier states (not just the nuclear weapon states) to halt proliferation, and as such carry a strong political message. Such a political message may well have an effect on the demand side of the equation as well. States that are keen to support non-proliferation efforts interpret export controls as support for their approach. Thus, the vast majority of states support export controls.

Difficulties arise when state have been denied access to important technology through legal means – often for reasons of concern over their intentions and proliferation potential. They then have choices: try to obtain the technology through other legal means, through illegal means, manufacture it themselves, or halt whatever project it was needed for (this latter option may be the reason behind the supplier refusing the export). There may have even been instances where the denial of technology increased the desire to acquire nuclear weapons, although it is unlikely that denial in itself would initiate a nuclear weapons programme. Denial of legitimate technology can lead to a sense of resentment and discrimination, particularly when that technology may be readily available to other countries. Such resentment can nurse a sense of isolation that may lead to paranoia – a fertile breeding ground for would-be proliferators.[4]

In general, although export controls may have an effect on the vigour with which a state might pursue a clandestine nuclear weapons programme, the desire for nuclear weapons would have had to have been established beforehand.

CONCLUSIONS

Export controls have their place in increasing the degree of difficulty with which countries and terrorist groups can obtain sensitive technologies. However, they cannot prevent a determined proliferator which has the technical and financial abilities to manufacture nuclear weapons. While export controls can delay proliferation, as time goes on and the technology for a simple atomic bomb becomes more and more obsolete, that delay will decrease due to the enhanced technical capabilities of states.

Export controls affect the supply side of the proliferation equation, whereas their effect on the demand side seems to be minimal. Determined proliferators have developed a nuclear weapons capability despite the imposition of export controls. None the less, it continues to be important for supplier states to control their exports of sensitive technology. To abdicate such responsibility would be to give a *carte blanche* political signal to parties interested in nuclear proliferation. It is also important for supplier countries themselves to be responsible in their export policies so that their domestic public recognizes that their government does not tacitly sanction nuclear proliferation and that it is not party to the spread of nuclear weapons.

To maintain credibility, export controls need to be kept up to date, and equipment that can be easily bought for domestic use in most developed countries (such as powerful desktop computers) could be removed from control lists to provide the lists with a sense of reality.

In order for export controls to work more effectively, increased attention could be paid to the demand side of the proliferation equation in addition to the supply side. Addressing the demand side could include measures to help solve the national, regional and global security concerns of potential proliferators, recognition of the inequality of the non-proliferation regimes with steps to reduce the inequality (such as nuclear disarmament measures), security assurances, and recognition of the prestige and status of potential proliferating nations.

REFERENCES

1   Available at: HYPERLINK http://www.acda.gov/factshee/exptcon/zangger.htm; <http://www.acda.gov/factshee/exptcon/zangger.htm>. Emily Bailey, Richard Guthrie, Darryl Howlett and John Simpson, *Treaties, Agreements and Other Relevant Documents*, Briefing Book, Vol. 2 (Washington, DC: Arms Control and Disarmament Agency, Programme for Promoting Nuclear Non-Proliferation, 1998).
2   Nuclear Suppliers Group, *Memorandum of Understanding Implementing Guidelines for Transfers of Nuclear-related Dual-use Equipment, Material and Related Technology* (Washington, DC: US Department of State 31 ILM, 1984, 3 April 1992).
3   Emily Bailey, Richard Guthrie, Darryl Howlett and John Simpson, *The Evolution of Nuclear Non Proliferation Regimes*, Briefing Book, Vol. 1 (Washington, DC: Programme for Promoting Nuclear Non-Proliferation, 1998).
4   See, for example, Vijay Sen Budhraj, 'The politics of transfer of nuclear technology: A case study of the Tarapur agreement', *Australian Outlook* (1984) (Sydney: Australian Institute of International Affairs), p.21.

9 Control of Smuggling in Nuclear Proliferation

Alexander DeVolpi

INTRODUCTION

A major and dangerous aspect of technology transfer is the potential transfer of nuclear technology from the former Soviet Union to non-nuclear weapons states. This enhances the potential for nuclear proliferation.

Because of this threat, a project to deter nuclear smuggling out of the Russian Federation was funded in fiscal years 1997 and 1998 under the auspices of the US Department of Energy (DOE) Office of Arms Control and Nonproliferation. Labelled the 'Second Line of Defense' (SLD), its agreed primary goal has been to analyse and assess alternative strategies for developing synergistic systems to simultaneously deter smuggling of nuclear weapons, explosives, drugs, and so on at international and domestic control points. This work was initially carried out at the Argonne National Laboratory, a multi-purpose DOE research and development facility located near Chicago, Illinois.

In the context of 'avoiding nuclear anarchy', which corresponds with the title of an influential Harvard Study Group book on 'containing the threat of loose Russian nuclear weapons and fissile material',[1] the DOE/Argonne goal has been to help the Russians improve their equipment to detect nuclear materials at their borders. The Argonne strategy has been compatible with the recommendations of Graham Allison and his co-authors for concerted efforts to 'reduce the nuclear leakage threat as much as possible as quickly as possible'. Argonne's analysis, experimentally verified by its Russian contractor, the Kurchatov Institute, concludes that under current procedures, meaningful quantities of nuclear materials could now be passed without detection through monitors at customs checkpoints. The technical means to detect nuclear smuggling are not in place in Russia. Therefore, this analysis justifies the alarm sounded by the Harvard study, which concluded that 'nuclear leakage is a major threat'.[2]

The Argonne finding confirm that standard nuclear material monitors are unable to detect the required amounts of weapons-grade uranium or plutonium when enclosed in medium-sized packages or luggage. Hand-held radiation meters are useless when nuclear materials are deliberately shielded inside luggage and packages that move across border control points, nor are most materials used in the production of nuclear weapons (heavy water, beryllium, lithium, and so on) detectable. Only the implementation of advanced technology can close the current gap between export controls and measurement capabilities.

Working in co-ordination with export control authorities who have fashioned a regime of legalized regulations, the SLD Project has been aimed at enforcement of restrictions on illicit trafficking in materials of national security concern. Particular emphasis has been placed on the use of Russian-made-off-the-shelf equipment augmented by additional advanced technology where needed. While the project goal is directed at control points either in the USA or the Russian Federation, almost all the emphasis so far has been on application of control to border points of Russia.

All weapons of mass destruction – particularly nuclear, radiological and chemical – are included in the scope of intercept technology and procedures, and other export-controlled items such as nuclear production-related materials are also encompassed. Because of the need to support missions of other governmental agencies and to make the SLD cost-effective, collateral detection goals include high explosives, drugs, firearms, precious metals and toxic chemicals. Of course, high explosives – when associated with nuclear, radiological or chemical agents – increase the inherent danger from illicit shipments, and thus become of more direct importance for detection.

Nuclear materials include fissile and fertile substances. (Fertile substances are not themselves fissile, but can be converted into fissile materials by neutron bombardment.) For the detection of these, there is a well-established field of technology: nuclear materials diagnostics. While plutonium is not too difficult to discover, highly enriched uranium (HEU) and natural uranium are essentially undetectable when surrounded by ordinary packing within containers. Also nearly impossible to find by non-destructive means are the ingredients mentioned above for producing nuclear weapons source materials.

THE SECOND LINE OF DEFENSE PROJECT

The national security concerns of the USA are stimuli for the SLD programme. These concerns are amply described in publications and policy statements, some of which are quoted below.

SLD's Rationale

Specific recent publications on nuclear smuggling give recognition and impetus to US government and DOE actions. In particular, the book *Avoiding Nuclear Anarchy* has played an influential role. Written by Allison and colleagues at Harvard's Center for Science and International Affairs, it is noteworthy for being based on consultations with members of Congress, with policy-makers, and with non-governmental organizations.

This comprehensive study warned that nuclear leakage was becoming a major threat to US national security. The authors specifically endorsed the need for equipment to combat illicit trafficking in nuclear materials from the former Soviet Union (FSU). They noticed inadequacies in detection of nuclear materials both at Russian borders and at US points of entry – inadequacies yet to be remedied.

The Harvard study recognized that detection of nuclear weapons or fissile materials is 'not easy'. Noting that drug dealers have shipped cocaine in lead ingots, the study group determined that passive detection of shielded HEU is 'impossible'. They remarked that HEU, shielded or not, would show up in X-rays in ways that are 'indistinguishable from any other dense materials, like lead, or steel'. The authors of *Avoiding Nuclear Anarchy* conclude: 'Passive sensors are unlikely to be the solution to the nuclear smuggling problem.' Instead, the authors point to a solution in the form of 'active sensors that bombard unshielded HEU with an active radiation source designed to induce fission; the resulting neutron emissions can be detected at ranges of a few feet'.[3] According to this thorough study, little effort and progress has occurred in nuclear export control enforcement, and specifically even less in developing equipment needed to detect nuclear materials at points of entry and exit.

SLD's Goals

This type of analysis, as well as those conducted by others, has guided Argonne's activities. Because of its close association and participation with non-governmental organizations in the late 1980s and early 1990s, Argonne staff members were at the forefront in recognizing some of the nuclear problems arising from the break-up of the Soviet Union. Close contacts and rapport with Russian scientists and officials led to Argonne's proposals to counter the threat of nuclear smuggling.

As a result, a requirements-driven strategy has unfolded, based on a systematic assessment of requirements and capabilities for the detection of smuggling. Input has been supplied by relevant Russian ministries and institutions, as well as former US officials. The shortcomings of existing institutions and technology in dealing with the threat of nuclear smuggling have been taken into account. Inadequacies of existing technology to meet detection goals were identified at an early stage of the SLD project. The programme was originally designed to assemble existing equipment for testing and demonstration, while spurring improvements in advanced technology that would compensate for any identified shortfalls. Various inadequacies have been widely noted in publications, particularly in the Harvard study, which emphasized the uselessness of passive monitors in detecting uranium.

Steps to improve nuclear materials security in the former Soviet Union have been described in *Arms Control Today*.[4] Because of 'thefts of such material during the period 1992–1995', the DOE nuclear materials protection, control and accounting (MPC&A) programme was strengthened and expanded. Nevertheless, according to James Doyle of the Los Alamos National Laboratory, 'large quantities of weapons-usable nuclear materials in Russia and the NIS [Newly Independent States] remain inadequately secured'. He mentions 'the chance that "insider" personnel [who have suffered economic difficulties] could be tempted to divert nuclear material for financial gain'.[5] Not realizing that such a programme was already underway, Doyle advises preparation of a Second Line of Defense, noting: 'Despite the great progress that has been made in improving nuclear materials security in Russia and the NIS the risk that nuclear materials could be stolen or diverted remains significant. [Nuclear] material may already be outside these sites and under the control of individuals or groups that will attempt to smuggle it across national boundaries.[6]

In addition to nuclear materials, other weapons of mass destruction and explosives are vulnerable to being smuggled. Moreover, while the MPC&A first line of defense is being installed in Russia for 150 facilities within 53 sites, different considerations apply to nuclear smuggling through Russian borders which have many miles of boundary and multiple control points.

A nine-point SLD strategy – driven by requirements suggested by US and Russian officials and analysts – has been drawn up (see below). This requirements-driven strategy aims to have a meaningful and timely impact on the threat of nuclear smuggling. The strategy avoids equipment enhancements that would have the appearance, without the performance, of helping joint counter-smuggling goals. The strategy would bolster export control regimes that reinforce non-proliferation.

THE SLD REQUIREMENTS-DRIVEN STRATEGY

An SLD strategy driven by nuclear material-detection requirements has guided the Argonne initiative. This strategy was derived from four sources:

1   DOE guidance on US national security concerns
2   publications urging action to reduce the risk that nuclear materials would be smuggled out of Russia
3   discussions and documents supplied by Russian officials
4   technical and policy analyses carried out by Argonne staff.

This SLD requirements-driven strategy differs from a proposed supply-driven strategy, where available equipment and funding would be furnished to Russian institutions without ensuring that smuggling will actually be detected. The requirements-driven strategy is export control-oriented, applying limited US resources to assist in reduction of specific Russian border checkpoint detection inadequacies.

Nine stages in the strategy have been identified:

1   matching requirements and capabilities
2   using available technologies
3   recognizing measurement constraints
4   identifying inadequacies in capabilities
5   compensating for inadequacies
6   using equipment synergistically
7   applying DOE laboratory experience
8   testing SLD systems
9   focusing on SLD goals.

Matching requirements and capabilities

Through correspondence, phone calls and on-site visits, specific requirements from Russian officials and technologists have been solicited and tabulated. This information has been incorporated in a database for the SLD programme. The requirements have been specified in terms of the types and quantities of materials giving rise to national security concerns, and their forms of conveyance through border checkpoints.

Russian Federation requirements are based on information from their customs officials and from other law-enforcement agencies. Present capabilities of Russian Federation Customs to detect smuggled goods have been assessed and included in the SLD database. In addition, attention has been given to technologies available in Russia, especially at nuclear institutions and facilities. The Russian nuclear agency *Minatom*, its laboratories and other Russian institutions have been forthcoming in providing information on available technologies. For nuclear materials, the equipment furnished under the MC&A programme has been used as a standard.[7] For other materials, such as high explosives, instruments developed for US and international aviation and customs security have been evaluated.

The SLD database can be used to match requirements with available instruments. Incorporated into the database are specifications for instrument performance, as well as the names of suppliers. US interests can be accommodated by providing equipment and improved technology that will compensate for existing detection inadequacies.

Using available technologies

Argonne's survey of material diagnostics technologies has followed two directions: existing, off-the-shelf equipment, much of it commercially available, and advanced equipment that has been developed and tested but is not in production. Selection of appropriate technologies has been based on the requirements, recognizing that not all detection conditions need be met immediately for all materials and sites.

While most non-nuclear equipment will not specifically detect the materials of national security concern, they are well developed and have well-defined applications at air and customs checkpoints. For example, because much harmless metal is contained in checked luggage, metal detectors – which can be very sensitive – are primarily limited to human traffic. Hence, X-rays are currently the primary means for imaging contents of containers. Although in widespread use for aviation security, all variations of X-ray systems (single-energy, dual-energy and CT-scanning) respond primarily to differences in atomic densities. In order to detect materials relevant to the SLD design goals, chemical-specific responses would be needed. It must be possible to differentiate nuclear materials from innocuous substance, explosives from ordinary chemical fillers, and chemical agents from toiletries.

Existing capabilities to detect unsanctioned commodities can be quantified by deterministic and stochastic modelling using published instrument parameters. Possible synergies can be exploited among instruments that use different principles of operation. As a result of Argonne's survey of instruments and the modelling of capabilities, the conceptual design of a synergistic SLD system has been developed. In order to evaluate its performance, a demonstration in Russia has been planned.

Recognizing measurement constraints

Certain constraints apply to the Second Line of Defense that are not present in other lines of defence. These constraints have to do with the type of object inspected, the primary responsibilities of customs, the time available, various operational and human factors and cost-effectiveness. One of the main measurement constraints that separates the SLD from other defences against smuggling is the diversity of sizes and types of containers and fillers that might surround illicit objects. Another significant difference is the larger number of materials that are considered illicit or are export-controlled.

In developing SLD technologies, all containers are assumed to be closed and sealed, allowing only non-invasive instruments as the primary technical means of inspection. While containers could be opened either randomly or routinely, this is a time-consuming and uncertain process. For example, a metallic object found by a luggage search might still not be recognizable as benign without more invasive inspection.

Because the internal components of containers might be sealed or tightly wrapped, vapour and particulate sensors would have very minimal roles, even for explosives and volatile chemicals. Such sensors would probably have no role in detection of nuclear materials. Because a terrorist might slip up once in a while in sealing explosive packages, a synergistic suite of detectors, vapour and particulate sensors could have a useful complementary function.

Russian Customs has selected the validation of shipping manifests as one of the most useful and necessary tools. Transit of shipments accompanied by declarations is usually approved on the basis of examination of the manifest. Lacking non-intrusive means of inspection, decisions would have to be based entirely on non-technical suspicions. Even when an object is opened for inspection (such as a shipment of radio pharmaceuticals), equipment must be on hand that is capable of validating the manifest and ensuring that the radioactivity is not masking an illegal shipment.

The time available for inspection by hand depends on the number of items in transit and the number of inspectors. Pressures will always exist for expediting and circumventing the inspection process. Hence, technical means of inspection, while less subject to human vagaries, will be able to devote only limited time to each object. In addition, there are enormous problems associated with the large geographic scope of smuggling pathways and the very limited number of stationed control points.

In addition to improving detection technology, other actions could enhance counter-smuggling effectiveness. The quality and training of customs officials is directly and indirectly important. In order to be effective, equipment must be engineered for human use in the field. Moreover, the cost of equipment that has to be distributed to many internal and external border checkpoints cannot be excessive.

Identifying inadequacies in capabilities

Gaps in information on capabilities exist largely because of the wide variety of materials and containers. While capabilities are well established for detection of materials that are not shielded or hermetically sealed, the capabilities of instruments are not well known when these materials are surrounded so as to block or obscure the contents. Some inherent limitations in materials diagnostic capabilities are known to exist. For example, weapons-grade uranium is easily shielded.

Proposals have been made by some of the US weapons laboratories to rapidly enhance Russian Customs capabilities by installing portal monitors at airports and other customs checkpoints. However, the Argonne studies find that such costly assistance would provide only a very small improvement in existing capabilities. The proposed minimal allowed quantity of HEU, which is 10 grams according to the SLD database assessment, would escape detection by almost any normal detector. Just 43 grams of lead and iron will reduce the sensitivity of standard MC&A portal monitors by a factor of more than 500. The amount of deliberate shielding required to mask uranium would be no larger than a marble or large ball-bearing.

Many containers include metallic objects, so the size and shape of a metallic ball is not likely, by itself, to arouse suspicion. Not enough heat would be produced in 10 grams of uranium to be detectable. The spontaneous neutron-emission rate is negligible. X-rays would not reveal anything more than an opaque object. Thus, no combination of standard instruments is likely to meet HEU-detection requirements.

Passive radiation monitors have been successfully applied to deter and detect nuclear materials, including HEU, at nuclear storage facilities where pedestrians and vehicles pass by controlled portals. However, at customs checkpoints, the much larger traffic in various packages and containers of many sizes is conducive to efforts to avoid detection.

Because of self-shielding, our computations indicate that kilogram quantities of HEU would probably escape detection by passive radiation-detection portal monitors. No means other than active neutronic interrogation would be practical to detect shielded HEU. Weapons-grade plutonium is more detectable, but portal monitors would not be able to meet the 1 gram detection requirements for plutonium when it is covered by specially designed shielding nested within larger containers. Equipment specifications admit that 40 grams of plutonium could be effectively shielded in a package of reasonable size and weight. Hence, no improvement in detection at Moscow's Sheretemeyvo airport should be expected if their existing passive radiation monitors were simply replaced with other passive portal monitors.

For illicit chemicals and explosives that are well sealed in standard containers, identifying and distinguishing them from innocuous substances requires chemical-specific non-destructive detection. The full range of nuclear production-related and export-controlled chemical elements – such as deuterium, lithium production-related and export-controlled chemical elements – such as deuterium, lithium and beryllium – are undetectable in almost any quantities, if even minimal concealment is used.

These inadequacies in technology have been recognized during the course of the SLD project, and measures have been proposed to remove these inadequacies. Both analytical and experimental measures can be undertaken. The most critical inadequacy is the inability of off-the-shelf technologies to detect significant quantities of weapons-grade materials, even in a single shipment.

Compensating for inadequacies

To compensate for the inadequacies identified in existing materials diagnostic technology, several technical directions have been considered. One of these advances is to continuously update an integrated SLD information database that would register the technical requirements and capabilities. This database could also be accessible on a real-time basis for comparing field data with contraband profiles.

In order to assist in defining practical limits, field demonstrations of existing and advanced technology could be carried out. For this purpose, at the direction of the DOE, a two-stage demonstration was designed and planned to take place in Russia. However, without equipment known to accomplish chemical-specific detection of the primary materials of national security concern, such a demonstration would be limited in value in gaining field experience applicable to advanced equipment.

A systematic programme of analysis, using deterministic and stochastic modelling, can greatly reduce the cost of instrument development and increase confidence in its application. In addition, equipment will need to be tailored to the specific control point and to the smuggling pathway. Specialized equipment that is not commercially available needs to be engineered for SLD applications. In particular, neutronic interrogation systems (with either a pulsed or continuous source) need to be adapted for SLD use. Neutronic interrogation would satisfy the requirements for chemical-specific identification, and advanced neutronic equipment would have a spatial resolution that could greatly assist in defining small objects. The primary technical means now lacking is an engineered high-energy neutronic probe. Either pulsed- or continuous-source 14 MeV neutrons could demonstrate the capabilities of such non-destructive interrogation. Both inelastic and elastic scattering are fundamental types of neutron interactions that yield chemical-specific information. Three-dimensional imaging of the elemental composition of the internal components of a container could be achieved by directional neutron sources using the associated-particle method.

When combined synergistically with other instruments, in particular with X-ray systems that give better spatial resolution, a powerful detection capability could be achieved. The case for fast-neutron probes is based on two major advantages over X-rays and other forms of interrogation: deep penetration into objects, and tell-tale reactions created in materials. Aside from psychological resistance, neutrons do have some difficulties, but they can be overcome: their direction cannot be easily channelled, and their energy cannot be conveniently varied.

The d(t,α)n reaction overcomes these difficulties. This is a neutron-producing charged-particle reaction that results when deuterons collide with tritium in a well-established commercialized neutron generator. The first important outcome is that mono-energetic 14 MeV neutrons with a known velocity are produced (5×10^7m/s). Second, the associated-particle technique, which is a specific means of producing the 14 MeV neutrons in a sealed tube, can be instrumented to recognize specific beam directions and divergences. Third, because the neutron flight begins exactly at the time it is generated, nanosecond time resolution is achieved. Fourth, detection of gamma rays and/or neutrons results in identifying the type of materials and two coordinates of location within the container. Fifth, the time of flight defines the third coordinate for the place of interaction in an object. Sixth, the energy of the detected gamma ray completes the identification of the specific chemical elements within the object. Thus, fast-neutron interrogation systems meet the requirements for detecting and identifying substances of national security concern.

The technology for taking advantage of the d(t,α)n reaction has been developed and tested in laboratory environments.[8] The pulsed fast-neutron generator has been advanced furthest, but it requires some form of collimation to obtain the required spatial resolution.[9] This entails either a large, expensive system or time-consuming single-pixel scanning. The latter approach would be useful in a Russian demonstration to confirm the physical principles and practicality.

The associated-particle sealed-tube neutron generator (APSTNG) embodies the requisite directional capabilities to carry out three-dimensional imaging of the chemical composition of sealed containers. Single-pixel (wide-beam) systems have been on the market for many years, and they have accumulated meaningful data.[10] Since promising results have been achieved, a complex but more efficient multi-pixel system has been under development at Argonne, having been previously supported by DOE research funding. In addition to having detailed wide-angle spatial resolution, the multi-pixel APSTNG is intended to establish durability, a high neutron flux, a longer lifetime and improved reliability. After considerable efforts to design and implement the multi-pixel alpha detector for the APSTNG, it has been delivered by the manufacturer to Argonne. The APSTNG is a sealed accelerator tube, so that external tritium contamination is avoided. Because its neutron production is efficiently harnessed, no radiation shielding is needed or used in laboratory measurements. With the new multi-pixel APSTNG, it should be possible to confirm the expected capabilities of chemical-specific identification and localization of a wide range of dangerous substances (such as nuclear materials explosives, and chemical agents) and other materials (such as nuclear-production-related items, firearms and drugs). Most importantly, this non-destructive identification of the contents of containers should be achievable at an acceptable throughput rate.

With regard to detection of nuclear materials, fast fission can be induced by 14-MeV neutrons. Fission is detectable by a distinctive multiplicity of neutrons and/or gamma rays. Fast neutrons have adequate cross-sections (probabilities) for inducing fission in all fissionable materials, so that even natural uranium and thorium are detectable. Many other systems that rely on neutron-induced fission are limited to detection of thermally fissile materials. Several international and domestic programmes have been making use of fast-neutron interrogation. Recognizing the limitations of passive radiation detection, some Russian institutes have been working on both pulsed and associated-particle systems for explosives and nuclear materials detection.

In the area of aviation security, attention has been given to neutronic interrogation. A newspaper article in early 1998 about airport monitors pointed out: 'Bomb detectors as efficient and reliable as x-ray machines do not exist.' Imperial College, London, is developing an inelastic- and elastic-scattering pulsed-neutron explosives detector for airports. Norwegian governmental and private organizations have been considering purchasing APSTNG systems for airport detection of explosives. The Safe Skies Project of the US Federal Aviation Agency (FAA) included APSTNG development, and an FAA-sponsored conference in Monterey, California, in 1988 had special sessions on neutronic interrogation. The US Department of Defense is funding pulsed-neutron sources for detecting buried explosives, such as landmines and unexploded munitions. The fast-neutron spectrometer aboard *Lunar Prospector* was recently used for detection of water frozen near the poles of the moon. These international activities reinforce the decision to deploy fast-neutron interrogation equipment for chemical-specific detection and identification of smuggled items.

Using equipment synergistically

Disparate items of equipment, when linked in a synergistic system, can gain advantages in overall performance and in system and operational engineering. The design of a synergistic system of appropriate components would be based on the realization that no single instrument can meet all the SLD requirements. However, in combination – with appropriate data fusion and interpretation – synergies between the instruments can be exploited.

Examples of non-nuclear equipment that can used for screening containers at checkpoints include metal and vapour detectors. Different types of X-ray systems have been the mainstay for inspection of checked luggage at airports. The most promising combination of equipment, acting synergistically, would be a combination of non-nuclear screening devices accompanied by an integrated fast-neutron interrogation and X-ray imaging system. Data accumulated from these instruments could be processed and compared on line with other information, including manifests and security profiles. The results could be made available to operators on a near-real-time basis.

Applying DOE laboratory experience

DOE laboratories have considerable expertise in detecting nuclear and other materials. Because of the prominence of nuclear materials in the SLD programme, the initial reaction is often to expect that technology from the nuclear materials safeguards programme would be applicable. Although some of it might be directly transferable, much of it is not, because of the different circumstances of application. Tables 9.1–9.3 define and compare highlights of the three lines of defence against smuggling.

The capabilities and differences evident in the three lines of defense against nuclear smuggling indicate that a co-operative programme among

Table 9.1   Functional goals of the First and Second Lines of Defense against proliferation

	First Line of Defense (safeguards at nuclear sites)	*Second Line of Defense (customs control points)*
Fissionable materials	x	x
Radioactive materials	x	x
Nuclear production-related materials		
Chemical materials		x
Collateral materials (drugs, etc.)		x
Facility protection	x	
Fence control	x	
Materials accounting	x	

laboratories could incorporate the most appropriate instruments in an SLD system. Argonne has had considerable experience in developing a comparable synergistic system for the US Department of Defense. The fast-neutron/gamma-ray 'hodoscope' has had over three decades of successful operation in material diagnostics.[11] The goal of the hodoscope, which is an elaborate multi-channel system, is the detection of fissile, fertile and other materials in nuclear reactor safety experiments. Although hodoscopes are intended as permanent installations at reactors, their design principles and experience are relevant to material diagnostics.

Besides their focus on nuclear materials, other useful features of hodoscope operation include their application to sealed containers, stationary and moving objects, time-critical resolution, and ample attention to operational and human factors. The containers that have been imaged (with millisecond time resolution of nuclear material movement) have thick walls and diverse fillings. The surrounding environment has been saturated with obstructions and with high background radiation. And, because of the expense and importance of nuclear safety experiments, extremely high reliability has been required, implemented and achieved.

The hodoscope at the TREAT test reactor in Idaho consists of three types of neutron and gamma-ray detector arrays, with 700 active detectors operating to achieve simultaneously time-resolved and spatial resolution. The hodoscope has to function synergistically with other instrumentation, such as monitors for power, temperature, pressure and flow. Requirements and

Table 9.2   Operational and organizational factors for the First, Second and Third Lines of Defense

	First Line of Defense	Second Line of Defense	Third Line of Defense
Basic role	Nuclear site safeguards programmes	Detection and identification	Forensics
General location	Designated internal facilities and gates	Export transit sites	Foreign destinations
Specific location of detection	Designated internal facilities and gates	Customs control points	Field and laboratories
Operators	Specialists	Officials	Technicians
Materials analysis methods	Non-destructive and chemical assays	Non-destructive assays	Chemical and non-destructive assays
Lead Russian agency	*Minatom*	*Minatom*	*Minatom*
Lead US agency	DOE	DOE	DOE
Implementing Russian agency	*Minatom*	Russian Customs	Federal Security Bureau

design-basis capability vary for time resolution from 160 microseconds to much longer, and for data collection from seconds to hours. Detailed time-resolved mass and spatial resolutions are achieved. With an integrated data collection system, both real-time and post-measurement data processing and analysis are performed.

A hodoscope based on the US design has been operating in France. Proposals have been made for constructing a hodoscope in Russia. Materials diagnostic spin-offs from hodoscopes have been applied to nuclear arms control and cargo-scanning measurements.[12]

Table 9.3 Technical means employed by the First and Second Lines of Defense

	First Line of Defense (safeguards at nuclear sites)	*Second Line of Defense (customs control points)*
All materials	Non-destructive and chemical assays	Non-destructive assays
Fissionable materials	Dedicated, special-purpose equipment for materials	General-purpose instrumentation for wide range of materials
Radioactive materials	Radiation detectors	Radiation detectors
Nuclear production-related materials	Not applicable	Advanced non-destructive assays
Chemical materials	Not applicable	Advanced non-destructive assays
Collateral materials	Not applicable	Advanced non-destructive assays
Facility protection	Intrusion sensors and response mechanisms	Not applicable
Fence control	Mechanical, electronic	Not applicable
Materials accounting	Record-keeping, inventories and auditing	Not applicable

Testing SLD systems

Experiments are necessary to validate performance under the range of conditions that make the Second Line of Defense different from other measures to deter nuclear smuggling. The experiments need to be carried out with appropriate material surrogates and with a standard set of containers. Laboratory experiments could assist in demonstrating the capabilities and limitations of equipment before moving to field testing. A productive phase in developing and fielding SLD equipment would be to establish a testbed where standard samples and containers would be used to evaluate performance against requirements. Argonne National Laboratory has proposed a simulation testbed that would carry out performance evaluations, as well as test equipment for engineering suitability.

Practical material surrogates are necessary for two reasons: some of the substances (such as explosives and chemical agents) are too dangerous to use in the field, and common standards are needed that can be exchanged and used for calibration. Other innocuous surrogates could be chosen as appropriate for X-ray machines and non-nuclear instruments. From the viewpoint of demonstrating the physics involved, the following surrogates would suffice when fast-neutron interrogation is the primary means of materials diagnostics:

1   Natural uranium could represent all fissile and fissionable materials.
2   Caesium-137 would be a stand-in for radioactive materials.
3   Explosive simulants developed by the US Federal Aviation Agency are designed to have the correct physical properties.
4   Chemical-agent simulants prepared by the US Army have safe compositions of the dominant elements.
5   Heavy water or beryllium would represent controlled articles used in the production of fissile materials.
6   An unloaded handgun would exemplify the class of illicit metallic objects.

By selecting proper surrogates, safe standards would be available to prove the efficacy of fast-neutron interrogation for all classes of materials of national security concern. By testing other instruments for their ability to detect and identify these surrogates, the capabilities and inadequacies of alternative and complementary instruments could be quantified.

Focusing on SLD goals

The overall goal of the Second Line of Defense project is to detect materials of national security concern – in particular nuclear materials, materials used in nuclear production, and weapons of mass destruction. In order to make such systems cost-effective, other collateral objects need to be recognized, including explosives, precious metals, firearms and drugs. Only a synergistic system comprised of non-nuclear and nuclear equipment would appear to meet these goals. Neutronic interrogation, coupled to high-resolution X-ray imaging, would be a central component.

In order to deter nuclear smuggling at the pace recommended by various governmental and non-governmental analysis, a rapid-analysis programme would be needed – one that includes SLD database improvements, predictive analysis refinements, advanced-technology upgrades and implementation, systems engineering and field demonstration.

DISCUSSION

While this chapter has mostly focused on technical requirements to control technology transfer through enforcement of export control, non-technical measures are essential as well. For example, international law-enforcement mechanisms need to be strengthened. Another limitation of the scope of this paper is that it concentrates on checkpoints established at borders. Because leakage of nuclear materials is often more likely to occur in places that are not well patrolled, remote border monitoring also needs to be effective.

Finally, the magnitude of the nuclear leakage threat is dependent on the amount of weapons-grade materials in existence. The demilitarization and disposal of weapons-grade nuclear materials would be another measure that would reduce the threat presented by potential circumvention of the barriers. This chapter has placed particular emphasis for the SLD programme on the need to develop a counter-smuggling strategy by systematically assessing requirements and capabilities. 'Rapid-enhancement' proposals to contribute to or finance off-the-shelf equipment have an appeal that would not be matched by solutions to the technical problems of detecting smuggled materials.

Based on the requirements that have been derived from surveys of publications, meetings with officials, calculations using sophisticated computer models and experiments at laboratories, the following recommendations are made for additional work to meet the goals of deterring smuggling in materials of national security concern. As indicated, the recommendations are derived from a requirements-driven strategy that aims to ensure meaningful and timely emplacement of equipment, procedures and regulations to meet non-proliferation goals.

Procedures and enforcement integration requires the following measures:

1   Refine national and multinational legal barriers, and integrate these with realistic enforcement goals.
2   Select technologies and procedures for enforcing export control regulations.
3   Generate a normalized set of requirements that satisfy both funding and implementing organizations.

Database improvements are required, including the following:

1   Continue the development of the SLD database, both in format and content.
2   Integrate field data from demonstrations with threat assessments from the SLD database.
3   Place the SLD database on line for universal use.
4   Link the SLD database with US and Russian export control requirements.
5   Create a restricted-access version of the SLD database for sensitive quantities.

Predictive analysis refinements include:

1   Continue computer modelling in order to interpolate and extrapolate experimental data and to improve detection systems.
2   Integrate computer modelling with the field data in order to improve decision-making.

Demonstrations are required for the following purposes:

1   Establish a testbed for evaluating, calibrating and normalizing equipment performance.
2   Collect neutronic-interaction data for surrogate illicit materials within containers filled with a diversity of legitimate goods.
3   Co-ordinate SLD activities among laboratories.

Technology upgrades include:

1   Develop a prototype and test and engineer an advanced neutronic probe.
2   Move advanced equipment into implementation.
3   Improve other technologies and procedures used at Russian and US Customs checkpoints.
4   Extend controls by enabling remote monitoring of borders outside of checkpoints.

SUMMARY

It is a daunting task to actually build an effective and durable materials security barrier at points of exit from nuclear weapons states and points of entry to other nations. Such barriers could be designed using legal and procedural controls to inhibit illicit transfers, and could then be implemented by enforcement provisions that include the technical ability to identify forbidden shipments.

The Second Line of Defense project has been a joint US-Russian endeavour to reinforce existing border controls around the Russian Federation. The goal is to deter illicit trafficking in materials of national security concern, particularly nuclear weapons-grade materials. In addition, materials that might be used in the production of fissile materials are to be detected. Collateral goals include detection and identification of other substances with destructive potential – such as explosives and chemical warfare agents – and substances that are smuggled for their commercial value, such as precious metals, radiopharmaceuticals and drugs.

A requirements-driven strategy should be based on evaluating current requirements and capabilities, and then devising the means to satisfy these requirements. Existing equipment available to customs inspectors is severely lacking in ability to detect and identify almost all of these critical materials. Off-the-shelf equipment that could be purchased is also unsuited for the task. Instead, there is need to develop advanced neutron-interrogation technology, supplemented by conventional X-ray and non-nuclear instruments; this combination, acting synergistically, would greatly increase the technical barriers to smuggling.

ACKNOWLEDGEMENTS

The work that is reported in this chapter was carried out at the Arms Control and Nonproliferation Program at the Argonne National Laboratory, Argonne, Illinois, under the auspices of the US Department of Energy. Many colleagues at Argonne and at scientific institutes in Russia made significant contributions. The Russian Federation's Minatom agency has also supported the project.

REFERENCES

1 Graham T. Allison, et al, *Avoiding Nuclear Anarchy: Containing the Threat of Loose Russian Nuclear Weapons and Fissile Material* (Cambridge, MA: MIT Press, 1996).
2 Ibid., p.72.
3 Ibid., p.67.
4 James E. Doyle, 'Improving nuclear materials security in the Former Soviet Union: Next steps for the MPC&A Program', *Arms Control Today*, Vol. 28, No. 2 (March 1998), pp.12–18.
5 Ibid., p.12.
6 Ibid., p.18.
7 L. Neymotin and V. Sviridova (eds), *MPC&A Instrumentation Catalog* (Long Island, NY, and Moscow: Brookhaven National Laboratory and All-Russian Research Institute of Automatics, 1998).

8   See, for example, E. Rhodes and C.E. Dickerman, 'Associated-particle sealed-tube neutron probe for nonintrusive inspection', in *Proceedings of the 14th International Conference on the Application of Accelerators in Research and Industry* (Denton, TX, 6–9 November 1996).

9   T. Gozani, 'Nondestructive testing', in *McGraw-Hill Encyclopedia of Science and Technology Yearbook* (New York: McGraw-Hill, 1996), p.229.

10   See, for example, E. Rhodes and C.E. Dickerman, 'Associated-particle sealed-tube neutron probe: Detection of explosives, contraband, and nuclear materials', in *Proceedings of the Moscow ISTC Symposium on Nuclear Physics Methods of Detecting Smuggled Explosives and Nuclear Materials* (Obninsk: 8–11 April 1996).

11   A. DeVolpi, 'Applications of cineradiography to nuclear-reactor safety studies', *Review of Scientific Instruments*, Vol. 55, No. 8 (1984), pp.1,197–220.

12   C.E. Dickerman, et al, 'Demonstration of gamma-ray transmission hodoscope technology for warhead counting', in *Argonne National Laboratory Report ANL/ACTV-90/4* (Argonne, IL: Argonne National Laboratory, 1990).

10 The Demand Side of CBW Proliferation

Jean Pascal Zanders

INTRODUCTION

Proliferation is the lateral spread of certain weapon systems from a place where they are available to another place where they have yet to be introduced. It is obviously a form of technology transfer. The proliferation process can take different forms. In some cases, entire weapon systems are transferred. In other cases, the recipient country acquires dual-use technology, knowledge, equipment or other commodities to set up a domestic armament programme. The notion of 'proliferation' is usually reserved for non-conventional – nuclear, biological, chemical – weapons and advanced weapon platforms such as aeroplanes and ballistic missiles. The first introduction of these weapons into a volatile region can seriously upset the military balance. If other countries counterbalance the move by also seeking such weaponry, a destabilizing arms race with consequences far outside the region in question may ensue.

In 1984, the United Nations confirmed for the first time that chemical weapons (CWs) were being used in the 1980–88 Iran–Iraq War, and many industrialized states became increasingly concerned that developing countries were incorporating chemical and biological weapons (CBWs) into their military arsenals. It soon became clear that many Western companies were supplying Iraq with the technological know-how, infrastructure and raw materials to produce CWs. At the time, most industrialized states had no laws or regulations prohibiting or restricting these transactions. Following the examples of the Co-ordinating Committee on Multilateral Export Controls (COCOM), the Zangger Committee and Nuclear Suppliers Group, several industrialized countries began to meet within the informal arrangement of the Australia Group to co-ordinate their national export control regulations with respect to CBWs. The participants in these meetings also agreed to common lists of goods which are critical to the manufacture of chemical or biological weapons.[1]

As no major trade in or direct transfer of chemical and biological weapons is known to exist, dual-use goods are central to the CBW proliferation mechanisms. This characteristic complicates the CBW proliferation issue. By definition, dual-use commodities have legitimate civilian applications, but they may also be used in armament programmes. Restricting the transfer of dual-use goods may thus hinder the development of the economic, technological, scientific and social base of the recipient country. The fear of such consequence has led several developing countries to express serious concern about multilateral export control arrangements by supplier states, such as the Australia Group. These countries also point to the 1972 Biological and Toxin Weapon Convention (BTWC) and the 1993 Chemical Weapons Convention (CWC), which prohibit the acquisition and possession of biological or chemical weapons, but also urge parties not to impede economic and technological development for purposes permitted by the conventions.

Since the end of the Cold War, the proliferation debate has shifted in some major ways. The 1990–91 Gulf War pitted the world's most advanced armed forces against a regional power armed with CBWs. Although Iraq's non-conventional capabilities had only a relatively minor impact on the conduct of military operations by the Allied Coalition, the war has had three long-lasting consequences. First, the United Nations Special Commission on Iraq (UNSCOM), tasked with the elimination of Iraq's CBWs and missiles, uncovered how much more advanced the respective programmes were than had been assumed, and how self-sufficient Iraq had become in the development and production of such weapons. Second, although the war speeded up the negotiation of the Chemical Weapons Convention, the problem UNSCOM has experienced in revealing the full extent of Iraq's CBW programmes has cast doubt on the effectiveness of verification mechanisms in disarmament treaties. This doubt has reinforced the conviction among certain industrialized countries that the biological, toxin and chemical weapons conventions must be supplemented with export controls to prevent further proliferation of chemical and biological weapons. Third, thousands of Coalition soldiers have experienced a variety of illnesses connected to service in the Gulf during the war. An increasing number of reports link the ailments to the many toxic chemicals present in the Kuwait theatre of operations, and the administration of preventive drugs and vaccinations to counter the effects of CBWs. The possible connection between low-level exposure to chemical or biological warfare agents and these illnesses cause concern that even the limited CBW capabilities of a small power can inflict long-term damage on the best-equipped forces. Several Western powers have subsequently launched major research and development and acquisition programmes to counter CBW threats in future interventions.

Another major development in the proliferation debate since the end of the Cold War is the acquisition of CBWs by terrorist and criminal organizations. In 1994 and 1995, a Japanese extremist religious cult, *Aum Shinrikyo*, released the nerve agent *sarin* in Matsumoto and Tokyo, killing 12 people and injuring thousands more. The Japanese investigation revealed that the sect also had an advanced biological weapons programme, and had even tested *anthrax* on unsuspecting citizens. Since 1990, the USA has suffered a series of major terrorist attacks, including the first major ones inside its borders, which caused a considerable psychological shock. Since the Japanese incidents, fear has increased that terrorist and criminal organizations have crossed a psychological barrier and may make CBWs their weapons of choice. While the chance of a war or a major terrorist attack in which CBWs are used remain relatively low, the consequence of a lack of preparation are extremely serious, and at present few Western governments feel that they can safely ignore the issue.

The history of the CBW proliferation issue and the policy responses by mainly the industrialized countries has kept the focus of analysis on the supply side of the proliferation process. Apart from some general observation about why certain states may wish to seek CBWs, the demand side is ignored in proliferation analyses. This leads to several misconceptions – for instance, the widespread belief that only so-called 'rogue states' pursue CBW programmes – that prevent deeper understanding of proliferation mechanisms, and thus preclude policy options which target the proliferation pressures in the recipient state. Disarmament treaties, for instance, target certain armament programmes inside a country because they aim to reduce particular arms categories to zero, and therefore oblige a party to abandon any aspiration to acquire the prohibited weapons. The non-proliferation imperative currently reduces disarmament to but one of several policy options to reverse the spread of CBWs.

This chapter investigates the various factors that induce a country to seek chemical or biological weapons, the domestic processes involved, and finally, how proliferation processes may be understood despite lack of insight into the decision-making practices of some countries.

IDENTIFYING THE PROLIFERATOR

Proliferation conjures up the image of an oil slick spreading steadily from a central point to cover an ever-widening area. There is a sense of inescapability: all parts of that area are equally affected, and unless a physical dam is erected, containment is impossible. This was essentially how the CBW proliferation threat was presented during the 1980s. In testimony to the US Congress, officials and policy analysts repeatedly stated that as many as 24 countries might be developing chemical weapons, a figure which had risen from less than half a dozen at the beginning of the decade. With over 100 countries possessing a chemical industry enabling them to produce CWs at short notice, it was feared that the number could rapidly increase further.[2] A similar assessment was made for biological weapons: as many as 100 countries might soon be able to manufacture biological warfare agents without outside help, because of the increasing availability of biotechnology and advanced expertise.[3]

In the 1990s, the threat assessment changed in two significant ways. First, the number of proliferators stabilized at around 20 states which have or may be developing nuclear, biological or chemical weapons, or their missile delivery systems.[4] As the figure now comprises four weapon categories, it is more difficult to isolate the CBW threat assessment. In November 1997, the US Department of Defense listed nine countries as having a CW programme in various stages of development, and seven as having a BW programme. Some countries, however, are conspicuously absent from these lists.[5] Second, rogue states have become the focus of proliferation threat assessments. A rogue state is generally undemocratic, geopolitically dissatisfied, hostile to Western interests, and unlikely to uphold widely accepted international norms of behaviour. This focus may help to explain the absence of some countries from recent CBW proliferation lists. More importantly, the qualification suggests that attempts to acquire chemical or biological weapons are closely correlated to the type of political regime. However, this is historically incorrect.[6]

Chemical warfare, as it is understood today, is a typical product of the second industrial revolution which took place in Western Europe and the USA at the end of the nineteenth century.[7] One characteristic of this industrial revolution was the increasingly utilitarian application of scientific principles driven by an economic rationale. The First World War provided the stimulus to apply this new scientific knowledge to warfare. For instance, most of the toxic chemicals used as warfare agents on the battlefields had been discovered decades earlier, but were not then considered by the military or scientists as new weapons of war. CW proliferation began as soon as those countries at the threshold of a CW capability moved to establish a research and production base dedicated to purposefully acquiring such weaponry, and erected a bureaucracy and decision-making procedures with the organization of CW employment and defence as their prime purpose. Since then, proliferation processes have taken on different forms.[8] Democracies as well as states with other forms of governance have had active CBW programmes, and have used or been prepared to use these weapons in war. During the past eight decades, the identity and the number of countries pursuing CBW capabilities has changed as new programmes were initiated or existing ones were abandoned. The intensity with which a CBW capability was pursued has varied greatly, so that even states with uninterrupted CBW programmes have at times found themselves totally unprepared to wage or defend against chemical or biological warfare. In fact, the form of governance has no bearing on *whether* a state will seek to acquire chemical or biological weapons, but does play a major role in *how* it will organize its CBW armament programmes.

In 1992–93, Russia, the UK and the USA released details of their respective proliferation threat assessments. This enables the comparative analysis of intelligence assessments of certain countries for the same period.[9] Table 10.1 summarizes the data on chemical, biological and missile programmes for the countries in the Middle East. At that time, Iraq was the only country which possessed a confirmed CW capability, and UNSCOM would not uncover the extent of Iraq's BW programme for another two years. The most striking feature of Table 10.1 is the consensus on the identity of the main proliferators. Specific details about possession, programmes or capabilities vary in some instances, and may reflect different intelligence data or the use of different sets of analytical criteria. The Russian Foreign Intelligence Service Report was singular in its claim that Israel has a CW capability. It also denied that Syria, a former Soviet client state, had biological weapons. Only one US intelligence source claimed that Saudi Arabia may possess chemical weapons.[10] Other agencies do not appear to have repeated the assertion since.

Of the 20 Middle Eastern states under consideration, all but six were systematically associated with CBW programmes in the three intelligence assessments. This may appear remarkable, especially in the light of the traditional arguments about why countries wish to acquire a CBW capability. These arguments are mostly linked to factors or developments external to the state seeking CBWs. Among the external causes often cited are: deterrence, self-defence (including the function of CBWs as force multipliers to offset the military superiority of a rival state), aggression and coercion. Status and regime survival are often advanced as internal causes for the proliferation of non-conventional weapons, but may be of lesser importance for CBWs, as strong international disapprobation tends to force governments to keep such programmes secret.[11] The focus on external causes follows mostly from the methodology: 'primarily a deductive exercise based upon the strategic requirements of particular states, the tactical needs of their armed forces, and the utility of chemical weaponry for Third World conflicts'.[12] The geographical limitations of the methodology thus exclude

Table 10.1   Comparison of intelligence assessments on the possession of chemical and biological weapons and missiles

	USA			UK			Russia		
	CW	*BW*	*M*	*CW*	*BW*	*M*	*CW*	*BW*	*M*
Algeria				N	N	Y	N	N	Y
Bahrain									
Egypt	?	P	Y	Y	N	Y	P	P	Y
Iran	Y	P	Y	Y	P	Y	Y	N/P	Y
Iraq	Y	P	Y	Y	Y	Y	Y	P	Y
Israel	?	P	Y	C	C	Y	Y	N	Y
Jordan				N	N	N			
Kuwait									
Lebanon									
Libya	Y	P	Y	P	P	Y	Y	P	Y
Mauritania									
Morocco									
Oman									
Quatar									
Saudi Arabia	?		Y						
Syria	Y	P	Y	Y	P	Y	Y	N	Y
Tunisia									
Turkey									
UAE									
The Yemen			Y						

Notes: CW = chemical weapons; BW = biological weapons; M = missiles. Y indicates statement of possession; N indicates statement of non-possession; ? indicates probable possession; P indicates a programme under way; C indicates capable; blank indicates no information given.

Sources: US assessments: Z.S. Davis, S.R. Bowman and R.D. Shuey, *The Proliferation of Nuclear, Chemical, and Biological Weapons and Missiles* (Washington DC: Congressional Research Service, Library of Congress, 8 April 1992). R.M. Gates, 'The proliferation of weapons of mass destruction and the intelligence community response', statement of Director of Central Intelligence to the US House of Representatives Committee on Banking, Finance, and Urban Affairs, 8 May 1992; J. Woolsey, testimony by Director of Central Intelligence, to the Senate Governmental Affairs Committee (24 February, 1993), and United States Information Service, 'Woolsey outlines US security concerns', testimony before the Senate Select Committee on Intelligence (Embassy of the United States of America: Brussels, 26 January 1994). UK assessments: J. Reed, *Defence Exports, Current Concerns*, Jane's Special Brief, No. 1 (Coulsdon, Surrey: Jane's Information Group, April 1993). Russian assessments: Foreign Intelligence Service of the Russian Federation, *A New Challenge After the Cold War: The Proliferation of Weapons of Mass Destruction*, released at a press conference, Moscow, 28 January, 1993 (translated from Russian by the Foreign Broadcast Information Service).

analysis of non-possessors of CBWs and of past CBW programmes in industrialized states.

Reference to the Third World may in itself be misleading. The states in Table 10.1 allegedly seeking chemical and biological warfare capabilities are among the most advanced and richest industrialized countries. However, other Middle Eastern states also belong to the group of rich and advanced developing countries. The six alleged proliferators allocate some of the highest percentages of gross national product (GNP) in the world to defence.[13] Other countries in Table 10.1 also rank among the global top 20 as regards defence expenditure. The hypothesis can be advanced that richer and more advanced developing countries, which reserve a large slice of their GNP for external security, support CBW armament programmes, but that these characteristics are not necessary indicative of an interest in acquiring chemical or biological weapons.

The supposition that chemical or biological weapons may offset geostrategic vulnerabilities is also weakly supported by evidence from the Middle East, where three important factors influence the balance of power: population size, economic strength, and territorial size and location. Governments may view chemical and biological warfare capabilities as a means of counterbalancing disadvantages in these areas. Yet all the countries systematically associated with CBWs, except Israel, have some of the largest populations in the Middle East. Chemical weapons in particular may also be attractive as a relatively easy or inexpensive way to deny enemy forces passage through relatively inaccessible or sparsely populated areas. None of the presumed possessor countries, with the exception of Israel, has a high population density. For the region, however, they still have some of the higher rates. Saudi Arabia, one of the most vulnerable states, has only an estimated 8 people/km^2. Only Oman has a lower population density. Chemical and biological weapons may have a high political value as strategic weapons, especially if they can be delivered by ballistic missiles with the range to target the major population centres of an opponent. A high rate of urbanization may thus imply a high degree of vulnerability, and could increase the attractiveness of a chemical or biological arsenal for deterrence or coercion. Egypt (about 45 per cent), Syria (about 50 per cent) and Iran (about 54 per cent) have some of the lower urbanization rates in the region, surpassing only Yemen (about 25 per cent) and Oman (about 9 per cent). The rates of urbanization in Iraq (about 73 per cent) and Israel (about 90 per cent) are comparable with those of the other states in Table 10.1. The data seem to suggest that several regimes are prepared to exploit an awareness of the relative strategic advantage offered by the high urbanization rates in other countries. However, considering that almost every country faces threats from many directions, the data fail to explain why other Middle Eastern states do not exploit this vulnerability of potential adversaries.

The argument of offsetting strategic disadvantages with chemical and biological weapons appears even more implausible if projected against the backdrop of the three major geopolitical cleavages in the Gulf region, namely between the member states of the Gulf Co-operation Council (GCC),[14] Iran and Iraq. The disparities between the countries with respect to territorial expanse on the one hand and population size and number of military personnel on the other are enormous. Between 1985 and 1992, the numerical imbalance in military personnel was greatly reduced as a consequence of troop reductions in Iran and Iraq and force increases in the GCC states.[15] Furthermore, since the Iran–Iraq War, the GCC states have acquired high-technology weaponry and missiles to counterbalance their numerical inferiority. Saudi Arabia, in particular, has given top priority to improving its air force – especially with the acquisition of AWACS early-warning aircraft – as this is the only one of its armed forces capable of patrolling or repelling an attack in the remote areas. The spending spree after the 1990–91 Gulf War reinforced this trend. The absence of an indigenous defence industry has always made the GCC members dependent on foreign suppliers of military technology and expertise. In other words, the countries that might gain the most from the force-multiplying effect of chemical or biological capabilities to compensate for their geographic and demographic disadvantages are, according to Table 10.1, not associated with CBW proliferation. Although they are acutely aware of their strategic vulnerabilities and consider ballistic missiles an appropriate part of their force posture, the GCC countries display little interest in chemical or biological weaponry.[16]

Since states within a geopolitical region all facing similar external threats, make different decisions regarding the acquisition of chemical or biological weapons, internal factors in each state must influence these decisions. With the exception of Israel, all the countries identified in Table 10.1 as seeking chemical and biological weapons have experienced revolution in the past five decades. Over half of the other countries are relatively stable, conservative monarchies, while the remainder have undergone abrupt changes of governance. If the alleged and confirmed possessors of CBWs are contrasted with the conservative monarchies, it may appear that the internal legitimization of the revolutionary governments through international prestige increases the incentive to acquire CBWs. However, international disapprobation counters this push factor for CBW armaments by forcing governments generally to keep the programme secret. The importance of the distinction between the revolutionary and conservative societies reaches deeper: the revolutions injected a Western-style – capitalist, communist or fascist – modernization ideology into the traditional societies. In the Gulf region, the dominance of oil-based industry and organic chemistry on which it is based may bring chemical weapons within reach, but these weapons none the less still present a formidable technological challenge. In addition, chemical and biological armaments require a societal culture that reflects that modernization ideology. The conservative Islamic Gulf monarchies – which, as far as is known, display little interest in such weaponry – strongly resist the influences of modernity. By contrast, the revolutions in the Arab republics were carried out by officers trained in industrialized states, and were based on concepts from industrialized societies. These revolutions theoretically increase the receptivity for a technologically complex form of warfare with CBWs. From this perspective, Israel is no longer the odd country out among the presumed possessors of CBWs: regarding education, technology and industry, it resembles the West in many respects. Iran is also less of an exception: before the revolution in 1979, the Shah had pushed to modernize the country in fundamental ways since the 1950s. Immediately after the revolution, the country faced the onslaught of modern technology in the 1980–88 Iran–Iraq War, and since then the secular rather than the religious pillar of power continues to drive the trend for modernization. Modernization is thus a key concept, because whoever seeks a CBW capability is developing leading-edge technology for that society.

The presumed possessors of chemical and biological weapons in the Middle East also share a fundamental dissatisfaction with the regional geopolitical status quo, which may be a further expression of the need for internal or external legitimacy for the regimes concerned. A global comparison between possessors and non-possessors of chemical weapons reveals the deeper meaning of this shared characteristic. As of January 1998, four countries are formally known to have CW stockpiles: India, Iraq, Russia (as successor state to the Soviet Union) and the USA.[17] The latter two countries apparently contradict the hypothesis that progress towards the so-called 'third industrial revolution' reduces the need for chemical weaponry, because high technology offers defence planners other options. However, India, Iraq, the Soviet Union and the USA have all had to meet any possible threat autonomously at every possible level of conflict. Before the Second World War, when they had to ensure their security independently, several second-tier European powers also maintained offensive CW programmes.[18] After 1945, they joined military alliances such as the North Atlantic Treaty Organization or the Warsaw Treaty Organization, whereby they made their security dependent on a large power. Consequently, they no longer had to meet each separate security contingency individually. A pertinent example is the UK, which ceased its autonomous offensive CW programme in 1956, and destroyed its last stocks of CW in 1959.[19] These moves coincided with its nuclear collaboration with the USA. For the Soviet Union and the USA, as leaders of their respective alliances, the post-war era caused little change to the principle of total self-sufficiency. India and Iraq, both with regional hegemonic ambitions, as evidenced by their respective nuclear weapon programmes, also seek military self-sufficiency. International isolation or the imposition of international sanctions against a country also reinforce the factor of total self-reliance. As the cases of Iran, Libya and South Africa illustrate, perceived military necessity and the possible symbolism of international defiance can easily overcome political and moral opposition.

The self-sufficiency explanation also appears to be valid in the Middle East, and again plausibly places Israel and the Arab states inside the group of proliferators. For years the conservative monarchies have made their security clearly dependent on the West, and on the USA in particular, and this dependency was confirmed after the 1990–91 Gulf War. Kuwait, for example, does not intend to acquire chemical weapons even in the event of a distinct threat or use of CWs, because it relies on the security guarantees extended by the USA.[20] In other words, the realization of the security deficit and the conscious choice of security dependency also plays a role in the political decision whether or not to proliferate.

UNDERSTANDING CBW PROLIFERATION FROM THE DEMAND SIDE

Motivations for arms acquisitions range from a state's uncertainty about its security in the international system to fundamental dissatisfaction with its geopolitical conditions. How states react to this environment depends less on external than on internal factors. All states face a complex combination of material, political and societal constraints which policy-makers must take into account when devising and implementing national security policies. These constraints also influence the nature of the weapons a state will acquire. According to the assimilation model, decision-makers must overcome these constraints if they wish to deploy a particular type of weaponry, and are consequently prepared to pay certain opportunity costs to achieve that goal.[21]

From the demand-side perspective, proliferation can be defined as follows:

1   Chemical or biological weapon proliferation occurs when a political entity decides to acquire a chemical or biological weapon capability where such a capability does not yet exist, provided this decision is followed by a chemical or biological weapon armament dynamic.[22]
2   Conversely, chemical or biological weapon deproliferation occurs as soon as the political commitment to that decision ceases to be renewed,

or if that political entity explicitly reverses such a decision. By defining proliferation as an armament dynamic in the recipient country, the process can be incorporated in the assimilation model of armament theory.

Assimilation is the process by which, for a particular weapon, weapon system or arms category, political and military imperatives, as constrained by the material base of the political entity, become reconciled with each other so that the weapon, weapon system or arms category become an integral part of current mainstream military doctrine. Any weapon, weapon system or arms category must consequently satisfy political as well as military imperatives. This presupposes the existence of a dual decision-making track: one in which military appraisals are primary, and one in which political considerations play the dominant role. The military track relates to those decisions taken by the military establishment to effect the military facet of the security policy of a political entity, including first and foremost the development and implementation of doctrine. The strategic planners will take into account external factors, such as the changing military threat, and internal factors, such as outputs of decisions on the political track. On the political track, overall policy decisions are taken regarding security and the means of implementing security policy. These may range from the formulation of a national security policy by the government and the parliamentary budget process to the expression of institutional interests inside and outside the armed forces, and inter- and intra-service rivalries within the military. As the military and political tracks interact, any decision, or set of decisions, not only influences future decisions on the same track, but also has ramifications for progress on the other track. A considerable level of tension may exist between both tracks, especially if actors on one track make demands which are irreconcilable with the basic goals or premises of the actors on the other track.

Any initial proposal for a particular type of weaponry envisages a particular end result. However, the weapon actually produced and deployed with the armed forces may differ significantly from the weapon originally anticipated. This variance between the original concept and the final product is the aggregate of all opportunity costs paid in the effort to achieve the original concept. As the proposed weapon system enters the decision process, multiple decision thresholds must be crossed. This process involves many discrete minor and major decisions at the various stages of the armament dynamic. To overcome any such threshold, an opportunity cost has to be paid. The opportunity cost may relate to a variety of issues, such as meeting environmental concerns, finding fiscal resources, convincing the military of the programme's utility, political opportunism, prioritizing allocation of resources to overcome technical difficulties, pressures for disarmament or from international humanitarian law, public opinion, and so on. Opportunity cost thus not only involves a monetary cost to overcome the obstacle, but also the expenditure of political capital to ensure continuation of the programme at a particular stage. Different times and circumstances may result in different opportunity costs to be paid for similar decisions at a comparable stage of the armament dynamic. For example, legal or moral objections to chemical and biological weapons will be easier to overcome in a country facing an acute external threat than in one located in a region with low-level tensions. Decisions against the armament dynamic are as crucial as those promoting its continuation: they will affect the outcome of the dynamic as a consequence of an increased variance between the original concept and the final product.

The nature of the thresholds is determined by intrinsic factors if they refer to the country's material base, and extrinsic ones if they relate to the environment in which the weapon system is being conceived. The country's material base constitutes a particularly important independent variable affecting decision-making on both the military and political tracks. It includes factors which can hardly, or not at all, be influenced by the decision processes within the time frame of the armament dynamic under consideration. It consists of a country's physical base – geographical position, territorial size, population size, natural resources, easy access to resources abroad – as well as the standard of the population's education, the level of scientific, technological and industrial development, economic strength, culture, and so on. In other words, all other factors being equal, differences between the material base of any two countries may account for the different characteristics and results of the respective outputs. Each of the intrinsic and extrinsic elements may raise or lower the opportunity cost for crossing the hurdle.

At the end of the armament process, the summation of all opportunity costs paid at each threshold will determine the final outcome of the weapon system. There are three theoretical outcomes:

1   The variance between the original concept and the final product is nil if the weapon system has been achieved as originally conceived without any (uncalculated) opportunity costs.
2   The variance is infinite if the aggregate opportunity cost is too high – if, for whatever reason or combination of reasons, the weapon system is not produced or deployed.
3   In most cases, the variance will lie between these two extremes, and will consequently reflect the deployed weapon system as the result of all opportunity costs paid.

These outcomes are valid only if it is accepted that the policy proposers will try to keep the variance as small as possible – an assumption which is embedded in the assimilation process described above.

FROM ARMAMENT TO PROLIFERATION ANALYSIS

It was noted above that countries, which had achieved the second industrial revolution, introduced chemical weapons to the battlefield in the First World War. After the Second World War, countries moving into the third industrial revolution gradually abandoned CWs, as nuclear and improved conventional weapons based on the emerging technologies were able to perform the battlefield task of CWs. The acquisition of a particular type of weapon technology or the incorporation of a certain mode of warfare into mainstream military doctrine can thus be correlated to the level of development of a political entity. As these levels of development can be compared and contrasted, comparative studies will identify the relevant thresholds, after which the means and methods of overcoming them can be investigated. Three different types of comparative analysis are possible.

Synchronic analysis between different political entities

Here, differences in political, social and economic organization manifest themselves in the type and height of the obstacles which will emerge during the armament dynamic. For instance, in a democracy greater energy must be invested in convincing parliamentary and extra-parliamentary opposition of the utility of the armament programme than in a dictatorship. A country with limited industrial development will have to seek greater help from abroad. Such comparisons will consequently reveal a series of thresholds, as well as their relative importance in function of the type of state structure, the material base and the political and military responses.

Diachronic analysis of analogous armament programmes in a single political entity

This comparative method will not only reveal differences in the development of the political and military organization of the country, but will also draw attention to the development of the material base (industry, technology, education, and so on) and its impact on historically comparable issues in the armament process.

Integration of the synchronic and diachronic approaches

This method enables the projection of a current armament programme in a developing country onto the history line of an industrialized state. Intersection occurs at some point on that history line. It represents the earlier stage of development of the industrialized state matching that of the developing country today. The comparison pertains to the material base of these countries, and thus highlights some major difficulties which the developing country would encounter when pursuing a particular capability. This third method lies at the heart of proliferation research.

The proliferation issue can be introduced into the assimilation model because, irrespective of time or place, an armament dynamic always faces thresholds which must be overcome one way or another if the proposed weaponry is to achieve operational deployment. The political culture, the security requirements and the composition of the material base of the proliferator define the characteristics of the barriers, and consequently the size of the opportunity costs to overcome them. Following the initial political decision to acquire a particular type of weaponry, the proliferator may encounter an important hurdle in its material base which cannot be solved by a mere decision on either the political or military track. This threshold therefore affects development on both tracks.

Elements, alone or in combination, that may play a role in defining the height of the threshold in the proliferating state are the scarcity of certain natural resources, lack of technical skills, insufficiently advanced education, an insufficient research and development or industrial base, and so on. Barring abandonment of the entire project, the political leadership has two basic options: either to develop the missing ingredients indigenously, or to seek them abroad. It may, of course, also opt for a combination of both. However, given the probable time frame within which the armament dynamic must be realized, importing the missing elements may be the only feasible and, in the short run, the cheapest option available. Especially if the dearth occurs in the physical base of the political entity, importation may be the only option. In other words, the decision and subsequent actions to seek certain ingredients abroad is but one way of structuring the armament dynamic of the political entity.

The hurdle to be surmounted because of the insufficiency in the material base is particularly high for a developing country seeking a chemical weapons capability, and important opportunity costs to overcome it may be envisaged. The size of these costs, however, will depend primarily on the extent of the preconditions for CBW armament that are already present in the political entity. The government, for instance, will have to consider the enormous financial implications a CBW project entails, as the economy of the country must be able to support the programme. As a consequence of the secrecy usually surrounding CBW programmes, and because of the requirement not to be distracted from the goal in spite of the many thresholds to be crossed, the government cannot count on economic offsets such as foreign direct investments in the domestic economy or technology transfers.

In addition, it is far from certain that embarking on a CW armament programme will enhance the security of the state. There may be significant international repercussions, especially if the importation of CBW-related materials affects the external security of other countries or if the dealings are undeniably illegal or contrary to international norms. Furthermore, the political entity makes itself dependent on foreign suppliers, and such sources can be shut off, affecting the overall security posture of the state.

The assimilation model can be used to study proliferation from the demand side. This follows from the presentation of proliferation as an armament dynamic within the proliferating state, rather than as a lateral diffusion of weapons technology from possessor to non-possessor states. Irrespective of the characteristics of the political entity, in the effort to assimilate chemical or biological weapons in mainstream military doctrine, the promoters of the armament dynamic will aim to keep the aggregate of opportunity costs as low as possible. Different times and places will generate similar hurdles, the height of which, however, may differ from political entity to political entity, or depend on the period under consideration. These differences lead to varying opportunity costs being paid to overcome the thresholds. The sum of these varying opportunity costs accounts for the potentially different outcomes of the dynamic in the countries under consideration.

Based on the premise that modern chemical warfare is an expression of a level of industrial and technological development comparable to that of the second industrial revolution, a current CW armament programme can be projected onto the history line of a Western industrialized state which once had such programmes. Political entities that have not yet achieved such a level of development are highly unlikely to have acquired a modern chemical weapons capability. Political entities that have surpassed this level of development tend to abandon an offensive chemical warfare posture, or display little interest if no such programmes had been active before. Other more advanced weapons can perform the same missions at least as efficiently, but do not entail the many objections to chemical warfare agents. Less information is available about biological weapon programmes, but at present no data suggest that the analytical framework is not applicable.[23]

The movement of chemical and biological warfare from the fringe towards the centre of mainstream military doctrine as part of the assimilation process will depend on how the political entity perceives and addresses its security deficit. An important variable in this respect is security dependency:

the degree to which a political entity is prepared to relinquish sovereignty over its security posture to another more powerful state. Reliance on a powerful custodian appears to function as a strong disincentive to offensive CBW programmes. In contrast, states seeking to maintain absolute sovereignty over their security posture perceive a greater utility for chemical or biological weapons. However, this is far from an absolute conclusion: the perceived utility of chemical or biological weapons diminishes fairly rapidly once alternative technologies materialize which can perform the same functions at least as efficiently, or which are more readily assimilated into mainstream military doctrine. Chemical and biological weapons have consequently always experienced great difficulty in maintaining a position close to mainstream military doctrine.

CONCLUSIONS

Chemical or biological weapons proliferation is far from the easy, automatic process which it is often depicted to be. From the demand-side perspective, the CBW armament dynamics in an industrialized society do not differ fundamentally from those in a developing country. The promoters will seek to structure the dynamic in such a way that the variance between the original plan and the final outcome remains as small as possible. They will consequently have to overcome thresholds whose nature and size depend on the social-political-economic fabric of the political entity involved.

Regarding proliferation, attention is specifically drawn to the material base of the proliferator. Important deficiencies in the material base may require the decision-makers to seek solutions from abroad: importation of certain commodities may be the fastest if not the cheapest way of structuring the domestic armament dynamic. While a certain level of technological, scientific and industrial development is a prerequisite for any political entity embarking on a domestic chemical or biological warfare armament project, importation – not the presence of the programme – testifies to the fact that the proliferator has not or cannot achieve a developmental stage present in the industrialized countries when they maintained similar programmes. In other words, the level of development of the proliferator may be expected to be lower than that of a Western industrialized country when it pursued the same generation of chemical or biological weapons for the first time. However, the proliferator today has the historical example and the knowledge about the properties of the agents and available production methods, so it need not research new agents; it can procure off-the-shelf technology to set up its own production base. In other words, it is able to choose its own time to commence a CW armament programme, and may decide to acquire an offensive CBW capability at a level of economic development lower than that of the most advanced belligerents in the First World War.

Taken together, these considerations suggest that the comparatively lower state of development at which chemical or biological weapon armament dynamics are activated may in fact be a standard feature among today's proliferators. The pursuit of such capabilities can thus be viewed as an expression of the limitations in the economic and industrial base of the political entity, which explains why such an armament dynamic still poses a formidable challenge. The aspirations can none the less be fulfilled because these limitations may be overcome through the importation of knowledge and technology widely available in the global markets. The fundamental dissatisfaction of these states with their geopolitical environment and result-ant expectation of war will lead them to adopt an economic policy that ensures the greatest possible degree of self-reliance and self-sufficiency. These states may thus have acquired several of the strategic industries necessary to sustain modern armed forces. The move towards a chemical or biological warfare capability may consequently fit into the long-term geo-political and industrialization strategies of these countries. However, if the level of development in the material base is indeed a key determinant in the structuring of the chemical or biological weapon armament dynamic, then related constraints may be expected to operate in other areas of armament as well. Indeed, the six countries systematically associated with CBW pro-grammes in Table 10.1 also display a remarkably high import-dependence for military hardware. They accounted for over 72 per cent of imports of major weapon systems in the Middle East between 1971 and 1990,[24] again demonstrating the failure to achieve or impossibility of achieving self-sufficiency in security matters. This increases the perceived security deficit, and strengthens the potential motivation to acquire non-conventional weap-ons to offset that security deficit.

ACKNOWLEDGEMENTS

Research for this paper was supported by a grant from the US Institute of Peace, Project USIP-027-97F 'Promoting Biological Weapons Disarma-ment'. The findings in this chapter are those of the author, and do not necessarily reflect the views of SIPRI or the US Institute of Peace.

REFERENCES

1   The history and functioning of the Australia Group is described in I. Anthony and J.P. Zanders, 'Multilateral security-related export controls', *SIPRI Yearbook 1998: Armaments, Disarmament and International Security* (Oxford: Oxford University Press, 1998), pp.386–94.

2   See, for example, the statement of William H. Webster, Director Central Intelligence Agency, before the Committee on Governmental Affairs, US Senate, 'Hearings on global spread of chemical and biological weapons: Assessing challenges and responses', 9 February 1989; statement of William F. Burns (Maj.-Gen. Ret.), Director US Arms Control and Disarmament Agency, to the Senate Governmental Affairs Committee, 10 February 1989; B. Roberts, oral statement to the Hearings on Chemical Weapons of the Senate Governmental Affairs Committee and the Permanent Subcommittee on Investigations, 10 February 1989. See also K.C. Bailey, *Doomsday Weapons in the Hands of Many: The Arms Control Challenge of the '90s* (Urbana and Chicago, IL: University of Illinois Press, 1990), p.58.

3   R.O. Spertzel, R.W. Wannemacher, C.D. Linden, D.R. Franz and W. Parker, *Global Proliferation: Dynamics, Acquisition Strategies, and Responses, Volume IV: Biological Weapons Proliferation*, Defense Nuclear Agency Technical Report No. DNA-TR-93-129-V4 (Alexandria, VA: Defense Nuclear Agency, December 1994), p.vi.

4   Counterproliferation Program Review Committee, *Counterproliferation: Chemical Biological Defense*, CPRC Annual Report to Congress (1997), Chapter 3, available at: <http://www.acq.osd.mil/cp/cprc97.htm>.

5   US Department of Defense, *Proliferation: Threat and Response* (Washington, DC: Department of Defense, November 1997), available at: <http://www.defenslink.mil/pubs/prolif97/>. As of 12 March 1998, it listed the following countries as having a chemical weapons programme: China, India, Iran, Iraq, North Korea, Libya, Pakistan, Russia and Syria. The countries which it listed as having a biological weapons programme are China, India, Iran, Iraq, North Korea, Pakistan and Russia. Libya was said to lack the scientific and technical base for a BW programme; Syria was said to possess the biotechnical infrastructure to support a BW programme. The absent countries were, notably, Egypt, Israel, South Korea and Taiwan, which were listed in the Office of Technology Assessment, *Proliferation of Weapons of Mass Destruction: Assessing the Risks*, OTA-ISC-559 (Washington, DC: US Government Printing Office, August 1993), pp.65–6. South Korea has meanwhile declared a CW production facility under the Chemical Weapons Convention; J.P. Zanders and J. Hart, 'Chemical and biological weapon developments and arms control', *SIPRI Yearbook 1998*, p.461.

6   India, the world's most populous democracy, was the first state to openly cross the nuclear weapon threshold since the entry into force of the 1968 Nuclear Non-Proliferation Treaty when it detonated five devices in May 1998. The US intelligence failure may in part be due to underlying assumptions about the behaviour of democracies.

7   The first industrial revolution took place in the middle of the nineteenth century, and was essentially characterized by the extensive mechanization of production processes in factories. In the second industrial revolution towards the end of the nineteenth century, discoveries in organic chemistry played a major role. Since the 1970s, a third industrial revolution is under way, driven by advancements in bio-technologies, electronics, information technologies, miniaturization, semiconductors, and so on.

8   J.P. Zanders, 'Towards understanding chemical warfare weapons proliferation', *Contemporary Security Policy*, Vol. 16, No. 1 (April 1995), pp.89–97.

9   Proliferation assessments of individual countries may have changed by the time of

publication. Several countries under consideration have meanwhile joined the Chemical Weapons Convention, and thus taken up the obligations to make full declarations about CW programmes and to allow international inspectors to visit facilities on their territory.

10 Testimony by Rear-Admiral Thomas Brooks, Director of Naval Intelligence, before the House Armed Services Committee in May 1991, as reported in Z.S. Davis, S.R. Bowman and R.D. Shuey, *The Proliferation of Nuclear, Chemical, and Biological Weapons and Missiles* (Washington, DC: Congressional Research Service, Library of Congress, 8 April 1992), p.11. In April 1989, the *Chicago Tribune* claimed that Saudi Arabia was one of the countries with access to the necessary resources for the manufacture of chemical weapons, but this report was immediately strongly denied by the Saudi authorities: 'SPA: Source denies chemical weapons charge', *SPA* (Riyadh), 6 April 1989 (in Arabic), Foreign Broadcast Information Service, *Daily Report – Near East and South Asia (FBIS-NES)*, FBIS-NES-89-066, 7 April 1989, p.19.

11 B. Roberts, *Weapons Proliferation and World Order* (The Hague: Kluwer Law International, 1996); p.114; E.M. Spiers, *Chemical and Biological Weapons: A Study of Proliferation* (Basingstoke: Macmillan, 1994), p.42. After Iraq admitted to possessing chemical weapons in 1988, President Saddam Hussein did refer to Iraq's so-called 'binary' chemical weapons as proof of the country's mastery of high technology to enhance the international standing of the regime and its survival in his speech of 1 April 1990, when he threatened to make fire eat up half of Israel. For discussion of the meaning of the speech, see J.P. Zanders, 'The chemical threat in Iraq's motives for the Kuwait invasion', in J.P. Zanders (ed.), *The 2nd Gulf War and the CBW Threat: Proceedings of the 3rd Annual Conference on Chemical Warfare* (Brussels: Interfacultair Overlegorgaan voor Vredesonderzoek, Vrije Universiteit Brussel, November 1995), pp.38–40. Through opaque communication – force posture, deployment patterns, refusal to sign the Chemical Weapons Convention and silence over international accusations of proliferation – Syria also signals its possible possession of chemical weapons.

12 Spiers, *Chemical and Biological Weapons*, p.42.

13 Comparisons have been calculated based on data in International Institute for Strategic Studies, *The Military Balance 1993–1994* (London: Brassey's, 1993).

14 The Gulf Co-operation Council comprises six states on the Arabian Peninsula: Bahrain, Kuwait, Oman, Qatar, Saudi Arabia and the United Arab Emirates.

15 In 1985, the GCC had a total of 137,800 troops under arms, compared with Iraq's 600,000 (excluding 425,000 personnel of the Popular Army) and Iran's 600,000 regular troops and Pasdaran (excluding several million paramilitaries). By the end of the war, Iran and Iraq's armed forces totalled 650,500 and 1,000,000 respectively, while the GCC countries had increased their armed forces to a total of 160,950. International Institute for Strategic Studies, *The Military Balance 1985–1986* (London: IISS, 1985) and *The Military Balance 1988–1989* (London: IISS, 1988).

16 In private discussions at the University of Kuwait in April 1994, all academics firmly rejected the CBW option to deter future aggression by Iraq or another Gulf power, but some wished for a nuclear capability, referring to the stability it introduced in European security relations.

17 J.P. Zanders and J. Hart, 'Chemical and biological weapon developments and arms control', *SIPRI Yearbook 1998: Armaments, Disarmament and International Security* (Oxford: Oxford University Press, 1998), pp.460–1.

18 See the contributions in T. Stock and K. Lohs (eds), *The Challenge of Old Chemical Munitions and Toxic Armament Wastes*, SIPRI Chemical & Biological Warfare Studies No. 16 (Oxford: Oxford University Press, 1997).

19   United Kingdom of Great Britain and Northern Ireland, 'Declaration of past activities relating to its former offensive chemical weapons programme', via the British Embassy, Stockholm, May 1997, p.2.

20   Private communication with the author by a senior official in the Ministry of Foreign Affairs, Kuwait City, April 1994. In other discussions (see note 16 above), Kuwaiti academics also discerned a clear positive correlation between their country being pro-Western and non-possession of chemical weapons.

21   J.P. Zanders, *Dynamics of Chemical Armament: Towards a Theory of Proliferation*, PhD Thesis in Political Science (Brussels: Vrije Universiteit Brussel, February 1996).

22   The armament dynamic embraces complex decision-making mechanisms in which numerous actors, positively influenced or constrained by environmental factors, participate. The arms acquisition process is thus the outcome of an aggregate of relevant individual decisions taken within a specified time frame. See Zanders, *Dynamics of Chemical Armament*, p.66.

23   The aspects which require further investigation are the impact of modern biotechnology and genetic engineering on the potential to devise more effective biological warfare agents, and the diffusion of this technology throughout the world as part of legitimate programmes to improve the quality of life.

24   Statistic computed from Y. Sayigh, *Arab Military Industry: Capability, Performance and Impact* (London: Brassey's, 1992), p.20, table 2.3.

11 Transparency and Verification

Bruce D. Larkin

INTRODUCTION

New technologies and new applications of technology, many derived from research and development undertaken to serve military objectives, have uses in the civilian sector. In practice, there is an interplay between military and civilian technology. Technology transfer has taken place in both directions. Defence procurement has always drawn from civilian technology, and post-Cold War budgeting creates added incentives to buy off-the-shelf technologies which have been created with a civilian market alone in mind. The drift also runs from military to civilian sectors. It does not require an explicit intent to convert and transfer: knowing that a technology exists, and seeing potential applications, breeds parallel and derivative civilian initiatives.

This chapter illustrates the interflow of technology streams between civilian and military purposes. It selects examples of transfers and exchanges which have a bearing on verification of weapon prohibition regimes, especially a regime of zero nuclear weapons (ZNW). As military reconnaissance technology moves into the civilian sphere, and is perfected and extended there, more states and even non-governmental entities gain access to technologies adaptable to arms control verification. This chapter aims to shed light on two issues:

1 How may military technologies and practices be adapted to disarmament goals, making disarmament politically more persuasive?
2 Is there a discernible trend in the pair 'concealment: discovery', suggesting the future efficacy of an abolition regime? Is there growing transparency? And if there is, will it help make ZNW work?

This chapter is concerned primarily with sensing objects which could be seen, on a clear way, from aircraft and satellites. However, it also notes new technologies for finding buried or concealed objects. Among places on the

Earth in which objects such as nuclear weapons could be hidden are the oceans and the ocean floor. Is there growing transparency even in this difficult region?

The chapter then introduces US and Russian steps to make Cold War oceanographic data available for environmental and other studies. The gist of the story is that the two countries gathered extensive data on the oceans because they saw them as operating areas for their own and their enemy's submarines. They are now making some of this data public, or in the case of the US MEDEA project, available to selected researchers who accept and meet US security clearance requirements. Some of this revealed data includes combined US and Russian data sets.

There are several disarmament implications of this readiness to open up data. It permits much greater clarity about an ambiguous medium – the oceans and seabed – in which nuclear weapons could be concealed. It reveals older sensing techniques, it provides a stronger basis on which to understand the presence and sensing requirements to monitor clandestine access to remote ocean depths, and it enables some Russian and US scientific participants to practise co-operation.

The chapter then reviews the repertoire of sensing technologies. Each exploits some correspondence between information in a sensible portion of the spectrum and the presence of physical objects or meaningful radiation. The aim in every case is to 'see' something significant. The chapter then exploits US Department of Defense public documents to identify possible future technologies which might, in the middle term, be adaptable to civilian purposes, and used to verify compliance with disarmament agreements.

The chapter then addresses technologies being marketed for wide dissemination in the civilian sector: video cameras, cellular phones, satellite phones and Global Positioning Systems. Some exploit technologies that were originally driven by military procurement. Although not intended as sensing or verification devices, or as stockpile monitors, they can be adapted and combined to complement monitoring systems, and significantly enhance their effect. UNSCOM cameras mounted at sensitive Iraqi sites, for example, permit continuous remote monitoring.

The final section considers conflicts and complementarities between civilian and military sectors. For example, the 'pull' for highly accurate Global Positioning System data by the civilian sector confronts the US Department of Defense perception that 'enemies' could turn accurate positioning information against US forces. On the other hand, the USA has published its intention to acquire commercial imagery for military purposes, when required, including views by 'high-resolution commercial sensors with enhanced spectral capabilities.[1]

DECLASSIFICATION DATA

Declassified Arctic data

Transparency extends not only to observations from space, but also to collecting data in environments which cannot be reached by conventional means. For example, there is no regular movement of civilian craft or scientific observers underneath the Arctic ice pack, but nuclear-powered submarine have manoeuvred under the ice since the 1950s.

The US Navy's Arctic Submarine Laboratory has released what *Scientific American* calls 'a wealth of information' gathered by those submarines from 1986. Their sea ice thickness measurements, made for navigation and defence purposes, may shed light on civilian studies of climate change.[2] This co-operation has been the subject of a well-illustrated *National Geographic* article,[3] and has been cited on the Website of the US Central Intelligence Agency as an example of participation in open, co-operative research activities.[4]

The MEDEA releases

In 1995, the US Navy issued a CD-ROM containing the report *Scientific Utility of Naval Environmental Data*, prepared by the Measurement of Earth Data for Environmental Data (MEDEA) group:

> In response to a request from Vice President (then Senator) Gore, an Environmental Task Force (ETF) was established in 1992 by the Director of Central Intelligence (DCI), including involvement by DoD and other agencies. The primary emphasis of the ETF study was on space-based systems and capabilities, including the National Technical Means. Some attention was paid to a variety of Navy systems and databases; however, that study did not encompass an in-depth examination of the full variety of the Navy's oceanographic data sets and capabilities.
>
> There was an opportunity to address those omissions with the formation of the MEDEA follow-on to the original ETF [Environmental Task Force] at the Navy's request, the present study was undertaken to examine the various classified databases, products, and capabilities of the Naval Meteorology and Oceanography Command ... The intention of this study was to determine the potential for unique and important environmental research arising from the use of existing classified Navy databases, and to prioritize these data for subsequent Navy declassification efforts. In addition, this study was to identify opportunities for collaboration between the civilian and Navy ocean science communities that could benefit both, and to suggest ways to obtain increased national benefits

from previous public investments in global data collection and modeling by the Navy.[5]

MEDEA has focused on ways in which data collected for naval purposes could be applied to issues in earth and environmental studies. The report illustrates collection from satellite, aircraft, ships, shore stations and deployed sensors, showing use of *Landsat*, GPS, P-3 *Lidar*, *Comsat* and *Argos*. Bathymetric data of the resolution released (with a 'high resolution' of 0.1 arc minutes or 200 m) appear too crude to be useful in finding concealed objects. There may be higher-resolution data in classified data sets. The MEDEA report authors also note a new technology for more precise modelling and identification of objects. They write:

> raw, side-scan sonar data offer an exciting opportunity to augment relatively low-resolution surveys ... The availability of high-resolution side-scan sonar coverage will have major implications for the oil and gas industry in mapping seafloor faults and fractures that control oil or gas seeps, produce subs[ea] freshwater springs, and contribute to slope instability. Hazards to navigation and existing pipelines can be much more accurately located, facilitating pipeline repair or ship routing around hazards.[6]

By contrast, however, detailed tracing of marine gravitational and magnetic fields may provide a backdrop against which the existence of unknown objects could be discerned. Whether that proves possible will depend on the degree to which effects are discernible. Writing generally about gravity measurements, the MEDEA report states:

> Over the past 30 years, as the need for precision in the positioning and navigation of space vehicles and other platforms increased, it became increasingly important to account for the slightest gravity variations. Thus, not only the general overall shape of the Earth (oblate spheroid) was needed to drive the gravity field, but ever smaller topographic features and geographic structures (which can cause gravity changes sufficient to affect critical instrumentation) had to be mapped.[7]

The MEDEA report continues:

> The Earth's shape has been measured with increasing accuracy ever since the development of earth-orbiting artificial satellites. In the 1980s the Navy measured the equipotential surface of the oceans, the geoid, very accurately using a satellite altimeter aboard *Geosat*. The geoid differs significantly from the reference ellipsoid. Not surprisingly, it is possible to derive the gravity field from the geoid; in particular, the product of the gravity field and the geoid anomaly in meters (i.e., the difference between the geoid and the reference ellipsoid) is

equal to the anomaly in the gravitational potential. The Navy's accurate measurement of the geoid thus allows a global inference of the associated gravity field.

The 'other platforms' to which MEDEA refers presumably include ballistic missiles. The US Navy also gathered geomagnetic data. Unlike the US-Russian co-operation on the Arctic, MEDEA's work involves a group of US scientists with security clearances assessing the scientific utility of selected disclosures of classified data. Jeffrey Richelson judges them 'a well-informed advocacy group in favor of further declassification'.[8] A table identifying specific scientific studies which could be undertaken using classified data, and the intelligence systems by which the corresponding data was collected, accompanies Jeffrey Richelson's discussion of MEDEA in *Scientific American*.[9]

REMOTE SENSING TECHNOLOGIES

The list of remote sensing technologies in Table 11.1 illustrates the range of methods available.

Many of these techniques have given rise to well-established, familiar methods of monitoring from aircraft and from space. US release of some 800,000 *Corona* images taken from satellites between 1960 and 1972 illustrates the quality of early photographs.[10] Today, anyone can view sample optical images on the Websites of commercial image-providers. No imagination is required to see how optical images could be used to spot large-scale activities and facilities. The reliance of UNSCOM on U-2 overflights to support ongoing monitoring activities in Iraq is widely reported.

Long-standing US use of satellites to detect the flash of a missile launch, originally designed to warn of missile attack, can also be adapted to reveal missile testing. A regime banning ballistic missiles except for space launch and research would doubtless include a prohibition of unacknowledged and unobserved tests, for which this detection would be a crucial element.

In addition to these familiar sources of information adaptable to arms control and disarmament purposes, other technologies and computational systems are being increasingly refined, and show promise for disarmament support. A study of ocean currents relies on a thousand automated stations which float with the current at depths up to 2,000 m, come to the surface, broadcast their position to a satellite, and then return to the deep.[11] Even the commercial proposal by a Russian submarine designer to build submarines as oil tankers to operate along Russia's northern coast, suggests further entry into a medium hitherto confined to military use.

Table 11.1 Remote sensing technologies

Technique	What it exploits	Advantages and drawbacks	Platforms (active/passive)
Optical observations	Reflection and emissions of light in the visual spectrum	Under the best conditions, great clarity, with resultant photos open to conventional photo-interpretation; hampered by night and clouds	Satellite, aircraft, ground; usually passive; can be active (floodlight, etc.); includes exploitation of low-level ambient light using photomultipliers
Infra-red detection	Reflections and emission of light in the infra-red region of the spectrum[1]	Can see through clouds and at night; poor resolution	Satellite, aircraft, ground; usually passive; can be active ('night vision' of illuminated objects)
Synthetic aperture radar (SAR)	Radar emissions	Permits precise measurement of distance to object[2]	Satellite, aircraft
Magnetic anomaly detection (MAD)	Distortions of the Earth's magnetic field	Useful to find large metal objects (submarines) from heights, or small objects by close examination	Passive
Gravitational anomaly detection	Distortions of the earth's gravitational field	Useful to find large masses	Passive
Radio-frequency monitoring	Data streams transmitted other than by secure optical fibre can be intercepted and read	Messages contain text significant in revealing activities and intentions; but 'eavesdropping' is ethically and practically troubled[3]	Passive, if accomplished by satellite or aircraft; active land-line intervention is also possible

Notes:
1 Under the rubric 'multispectral thermal imaging', heat given off is being widely exploited for environmental monitoring.
2 Results are averaged over an area, so small objects may not be distinguished if radar source is distant.
3 Probably accessible on a large scale only to major governments. Even they then must deal with the fact that sensitive materials are encrypted.

Airborne Synthetic Aperture Radar

An example of what can be shown by synthetic aperture radar (SAR) mounted on an aircraft is offered by the US National Aeronautics and Space Administration. Radar was used to find hitherto concealed archaeological remains at Angkor Wat. Dr Anthony Freeman, a radar specialist at the US Jet Propulsion Laboratory collaborating with archaeologists working on the site, said of the images: 'The radar image make apparent many features that are not readily identifiable on the ground. We can see differences in vegetation structure and some features that are obscured by vegetation cover.'[12] An accompanying graphic shows a 2.9 m resolution image.[13] Both applications to arms control and limitations are evident from the example.

Motion-detection

US plans to develop and orbit two synthetic aperture radar/ground moving target indicator (SAR/GMTI) satellites suggest that methods have found to achieve effective high resolutions.[14] Synthetic aperture radar works by measuring very precisely the time required to bounce an active signal off a surface. It is used, for example, to measure the 'height' of the ocean. One potentially limiting factor is the wavelength of the signal, but the current SAR systems permit discriminations of about 1.5 cm. Descriptions of early space-based use of synthetic aperture radar suggested that the result was an average of signals returned from a fairly large area, perhaps the size of a field, so that the method's spectacular success in vertical discrimination was not matched on the horizontal scale.

Presumably, the SAR/GMTI system exploits the fact that a moving object seen from the side is changing its distance from the sensor over time. Computer processing of successive signals reflected from a given small region, exploiting both redundant time information and Doppler effects, might be able to extract from the signals greater horizontal resolution. Under favourable conditions, the system might exploit simultaneous optical coverage of that region. In principle, however, a device which could not identify a moving missile launcher, such as a land-mobile ICBM, at night would be of little military value, so guesswork suggests the projected system would be designed to work at night, and therefore would not be dependent on visible light.

Thermal imaging

In March 1998, the US Department of Energy displayed a 'multispectral thermal imager' designed to extract information from radiated heat. From the description given, it appears able to distinguish intensity of emission, and characteristic fingerprints of radiation emitted by known chemicals. The version displayed was designed to be deployed in space, testing emissions from a height of (say) 600 km, but can also be aircraft-deployed.[15]

Detection from space of unavoidable or inadvertent chemical emissions can point to sites engaged in activities violating chemical, biological or nuclear bans. Effective use for arms control purposes requires identifying chemicals which are good discriminators and are unlikely to be produced in commercial activities, and mapping known commercial sources of poorer discriminator chemicals. Of course, a scheme of deliberate deception will seek to collocate barred facilities with legal operations, and to mask salient emissions, thereby presenting more difficulties for the analyst.

Magnetic anomaly detection

Hand-held magnetic anomaly detectors are used to locate archaeological objects up to a few metres underground. For example, Sheldon Breiner employs a portable $25,000 caesium magnetometer to search for Olmec objects at sites in Mexico. An account of his work cites 'weapons detection' as another use for this technology.[16]

SENSING PLANS OF THE USA AND UK

US defence planning is spelled out in a number of public documents, including a *Quadrennial Defense Review* (1997), the report *Joint Vision 2010*, and the *1998 Annual Defense Report*.[17] The latter report from the US Secretary of Defense makes clear that space assets will form a key element in future operations: 'DoD is moving into the information age and towards a totally integrated battlespace, where communications and intelligence space systems are no longer viewed as solely supporting capabilities to the warfighter, but as instruments of combat'.[18]

Not only would space figure in fighting wars on the Earth's surface and oceans, as in the past, but space itself is described as a region in which the USA has assets, and therefore a military interest:

Dependence on space force for military operations, as well as for civilian and commercial uses, is growing. The space C4ISR infrastructure, including terrestrial applications technologies, is expected to contribute tens of billions of dollars to the U.S. economy and may grow to hundreds of billions by 2000. During the next ten years, as many as 1,200–1,500 satellites may be launched – most will be built in the United States, and 30 percent will likely be launched by U.S. flag carriers. DoD recognizes these strategic imperatives and will assure free access to and use of space to support U.S. national security and economic interests.[19]

Nothing suggests the flavour and extent of the Department of Defense's intentions more clearly than its own words, including Global Command and Control Systems (GCCS) and Global Command and Control System-Top Secret (GCCS-T). These portray a blanket, extendable system in which reconnaissance and surveillance is integrated with military operations, both conventional and nuclear. 'The strategic vision for command, control, communications, computers, intelligence, surveillance and reconnaissance (C4ISR) is to provide capabilities that enable forces to generate, use and share the information necessary to survive and succeed on every mission. Major accomplishments in all areas of C4ISR bring the US Department of Defense closer to achieving this vision':

Information superiority provides the capability to collect, process, and disseminate an uninterrupted flow of information while exploiting or denying an adversary's ability to do the same. It includes comprehensive knowledge of the battlespace, including the status and intentions of both adversary and friendly forces. The Quadrennial Defense Review (QDR) identified information superiority as the backbone of military innovation, and noted that the Revolution in Military Affairs centers on developing the improved information and command and control capabilities needed to significantly enhance joint operations.[20]

They intend to make observations from space and from aircraft, to make what they see useful, and to put it in the hands of US forces so it can be used immediately to select and execute military manoeuvres and strikes. Moreover, they will block – or even turn to their advantage – their enemy's ability to sense and process. They wish to see all, but leave their enemy blind. It is not just 'processed' information which is to go to fighting units. The entire structure of command – the hierarchy through which reports travel up and orders travel down – is to be embodied in computer-controlled networks. Through these networks, commanders will tap intelligence. The network will come to include even the SIOP (Single Integrated Operational Plan), which governs all targeting with nuclear weapons.

COMMAND AND CONTROL

The *1998 Strategic Defense Review* says:

> Command and control (C^2) systems provide the means to effectively execute nuclear, conventional, and special operations. The Global Command and Control Systems (GCCS), which replaced the World Wide Military Command and Control System, provides nearly 700 locations with its secret level functionality and increased capability. GCCS provides an enhanced common operational picture, force status, intelligence support, enemy order of battle, related facility information, and air tasking orders. In 1998, GCCS Version 3.0 will provide imagery, meteorological, and oceanographic data. GCCS Top Secret (GCCS-T) provides a top secret infrastructure for C^2 throughout the force deployment cycle. When completed in mid-1998, GCCS-T Version 2.2 will add nuclear Single Integrated Operational Plan capabilities and a top secret (including special intelligence) common operational picture. GCCS and GCCS-T improvements in 1999 will further add sensitive compartmented information, increase user sites, and improve performance and reliability. DoD will evolve toward more integrated and interoperable battle management systems through continued deployment of GCCS below the joint command level and into operational units.[21]

The promise of improved 'reliability' in 1999 suggests that the system introduced displayed lapses in reliability. This is not surprising, as complex systems typically do so. Published analyses by the US DoD Office of Test and Evaluation confirm a healthy scepticism that all will work as planned.

At another level, airborne reconnaissance will be integrated with satellite data:

> To increase interoperability, the National Reconnaissance Office and the Defense Airborne Reconnaissance Office are developing complementary space and airborne surveillance and reconnaissance systems. Joint Signals Intelligence (SIGINT) Avionics Family (JSAF) sensor equipment will not only provide increased performance, interoperability, and commonality across the airborne reconnaissance fleet, but also allow interoperability with satellite systems.[22]

This will be augmented by a Moving Target Indicator (MTI) system. They will be able to 'see' what is going on better ('situational awareness').

Perhaps, with close airborne surveillance, it will be possible to discriminate movements of a handful of personnel:

> Manned airborne surveillance and reconnaissance assets are developing better situational awareness by using enhanced and modernized capabilities, such as Moving Target Indicator (MTI) and JSAF. In addition to the Joint Surveillance

Target Attack Radar System (JSTARS), the most robust and capable example of MTI surveillance, MIT capabilities have migrated to the U-2 and the Airborne Reconnaissance Low. While U-2's improved MIT-capable radar will begin delivery in FY 1998, both the RC-135 RIVET JOINT and EP-3 aircraft are completing other major upgrade programs and will begin transitioning to JSAF in FY 1999. JSAF equipment can be used not only in manned signals intelligence platforms, but also in UAVs, pending their adoption of the signals intelligence mission.[23]

Underlying the use of space assets is the expectation of information warfare in all operations. The US Department of Defense envisages that the USA:

> must have information superiority: the capability to collect, process, and disseminate an uninterrupted flow of information while exploiting or denying an adversary's ability to do the same. ... Information superiority will require both offensive and defensive information warfare (IW). Offensive information warfare will degrade or exploit an adversary's collection or use of information. It will include both traditional methods, such as a precision attack to destroy an adversary's command and control capability, as well as nontraditional methods such as electronic intrusion into an information and control network to convince, confuse, or deceive enemy military decision makers.
>
> There should be no misunderstanding that our effort to achieve and maintain information superiority will also invite resourceful enemy attacks on our information systems. Defensive information warfare to protect our ability to conduct information operations will be one of our biggest challenges in the period ahead. Traditional defensive IW operations include physical security measures and encryption. Nontraditional actions will range from antivirus protection to innovative methods of secure data transmision. In addition, increased strategic level programs will be required in this critical area.[24]

To the extent that the Department of Defense commits to the network as its means of carrying on military operations, it must make that network secure, redundant and survivable. With this in mind, it is understandable that the DoD considers mobilizing alternative civilian assets in time of need.

US military use of commercially available satellite data

It has been widely reported that the USA used commercial imagery from the French SPOT satellite during the Gulf War. SPOT provided coverage of larger areas than the more focused satellite reconnaissance undertaken by the USA itself. A small but growing number of vendors are offering satellite imagery for sale. How has the USA responded to this?

On the one hand, the USA reserves the right to limit sales of images by US firms, should it decide that barring access is required by US national security. In July 1998, it exercised that authority to ban one-metre resolution photographs of Israel (discussed below). On the one hand, it is laying plans to access commercial imagery itself, a mission entrusted to the National Imagery and Mapping Agency (NIMA):

> To meet long-term requirements, the National Reconnaissance Office has launched initiatives to revolutionize collection technologies used in space. NIMA acquires commercial imagery from multiple vendors for both geospatial production and peacetime and crisis applications. NIMA will also acquire unclassified imagery from new high-resolution commercial sensors with enhanced spectral capabilities. A joint government/industry team has been established to identify the best acquisition approach for the future. NIMA will migrate existing production systems to a more sustainable and flexible open architecture, and is shifting from predominantly hardcopy production, storage, and distribution to digital capability.[25]

UK plans

Other militaries are, of course, alert to the US emphasis on information technology, and have concluded as a result of their own analyses that they must integrate new technologies into their military planning more closely. What are UK plans? It is not known exactly how the USA and UK share information, but it is thought that the major UK listening centre GCHQ (Government Communications Headquarters) generates communications intelligence which the UK shares with the USA. This case suggests a tough question for disarmament verifiers: would states be significantly more able to satisfy themselves that nuclear disarmament promises were being kept if they could listen in on global communications? And would they find it politically and ethically correct to do so? It is understood that some portion of the raw material for communications intelligence is acquired by satellites which can gather microwave and other transmissions between points on the earth's surface, and between ground stations and communications satellites. Do champions of disarmament want to do that? Could they afford the cost? A clue to the importance nuclear weapon states attach to communications intelligence is the 1998 promotion of the head of GCHQ to the post of Permanent Secretary of the Ministry of Defence, the ministry's senior civil servant.[26] A decade earlier, that post was held by a man who made his name in nuclear policy.

The UK's 1998 *Strategic Defence Review* makes clear that it is following some of the same trends which animate the USA, and like the USA recognizes that there are increasingly capable systems available in the civilian sector:

Technological and social change will also open up broader possibilities which will have a profound effect on our future security. Many of these developments will be double-edged, bringing new vulnerabilities as well as opportunities. They include new ways of fighting such as information warfare (which attacks through the computer systems on which both our forces and civilian society increasingly depend); greater pressures on operational decisions (*instant media reporting* from both sides of the front line); the wider spread of technologies which may be used against us (such as biological weapons); and *highly sophisticated civilian capabilities that will be readily available both for us and potential adversaries*. And where we (and our Allies) exploit technology to strengthen our existing superiority in conventional weapons, our potential adversaries may choose to adopt alternative weapons and unconventional (or 'asymmetric') strategies, perhaps attacking us through vulnerabilities in our open civilian societies [*emphasis added*].[27]

On the specific question of arms control, the *Strategic Defence Review* explicitly cites an 'Open Skies' element:

The new Mission covers arms control, non-proliferation and related security building measures; our Outreach programme in Eastern Europe; and wider military assistance and training for overseas countries. We intend to improve the effectiveness of our existing activities and increase our effort in all these areas including ... an enhanced arms control programme incorporating an improved 'Open Skies' capability to monitor arms control agreements and additional training in arms control inspection techniques ...[28]

The important point to observe is that the UK Ministry of Defence envisions the public (and other governments) having capabilities which in some important respects are similar to the capabilities which the UK can deploy. It is likely that sensing capabilities will play a part in the designs for verifying nuclear weapons reduction and elimination which are spelled out in the *Strategic Defence Review*:

Verification of arms control and non-proliferation agreements is critical to their effectiveness, and has therefore been examined in the Review. It has traditionally been an issue on which Britain has made a substantial contribution. Over time we have developed particular expertise in the nuclear field in the monitoring of fissile materials, particularly through our involvement in the development of the IAEA's safeguards system, and in monitoring of nuclear tests. The Government intends to maintain these strengths, which will be important in implementing the Comprehensive Nuclear Test Ban Treaty and in negotiating a Fissile Material Cut-Off Treaty.
 ... But Britain has only a very limited capability at present to verify the reduction and elimination of nuclear weapons. A programme is therefore being set in hand to develop expertise in this area, drawing in particular on the skills of

specialists at the Atomic Weapons Establishment. A small team will be established to consider technologies, skills and techniques, and to identify what is already available to us in the United Kingdom. The Government will consider how to take this programme forward in the light of the team's interim conclusions. The aim is to ensure that, when the time comes for the inclusion of British nuclear weapons in multilateral negotiations, we will have a significant national capability to contribute to the verification process.

That last paragraph is significant because it records the first step by any of the five 'declared' nuclear weapon states to plan for denuclearization. All five have spoken of denuclearization, making commitments for some indeterminate future, and Mikhail Gorbachev offered a timetable in January 1986, but this is their first known step to charter personnel in government service to define practical measures in support of a ZNW regime. Many of the methods which could be employed would not require remote sensing, but others would.[29]

COMPLEMENTARY TELECOMMUNICATIONS AND GLOBAL POSITIONING

The 1946 Acheson-Lilienthal Report sketched simply but clearly a system of movement among nuclear specialists from laboratory to laboratory.[30] In this way, no laboratory would be without nationals of other states, and physicists would have opportunities to discuss their work abroad. The authors of the report anticipated that this movement would greatly increase the likelihood that any violation of the system of controlling 'dangerous' activities would be exposed. And that, in turn, would deter violation.

Today, analysts speak of 'societal verification'. A nuclear disarmament regime could include a call on all in participating states to expose violation, or suspected violation. Mordechai Vanunu's revelations of the Israeli nuclear programme illustrate the role individuals can play. Communication by telephone and the Internet provides society with ready means by which suspected violations can be revealed.

Moreover, new technologies permit novel combinations. Cameras are already used to monitor a site in real time, broadcasting what they see to a remote control centre. An object can be designed to give an alarm if modified – for example, if removed from a weapon system – and can, by exploiting the Global Positioning System, broadcast its own location. Physical enclosures can be monitored. Personnel serving as monitors can be in complete video and audio contact with supervisors, wherever they may be. The technical prerequisites for a complex fabric of assurances are already in hand.

CONFLICTS AND COMPLEMENTARITIES BETWEEN THE CIVILIAN AND MILITARY SECTORS

The Global Positioning System

There is no better example of incompatibilities between military and civilian requirements than controversies around the Global Positioning System (GPS). The USA deployed GPS with built-in errors, and reserves the capacity to turn GPS off, although it has promised not to do so.

Users who wanted only to approximate location were satisfied by the GPS accuracies initially provided, but any user whose requirements were more precise was denied that greater precision, on the grounds that a truly precise system could be used by an enemy to target military assets. The USA has addressed this by undertaking to 'identify a second coded civilian GPS signal and to develop a plan for providing the signal'. In the mean time, the USA agrees not to alter the GPS coded signal. This will 'assist civilian users in their constant quest for greater accuracy'.[31] The tension between civilian and military imperatives led President Clinton to create an Interagency GPS Executive Board, charged with reconciling 'global security' and 'economic' objectives.

An even larger question for civilian users is whether the USA would, on its own judgement, preclude their access to the system altogether. There are US plans for 'navigation warfare':

> Since the GPS has significantly military utility, and since it is in the best interest of the United States to prevent the hostile use of the system against U.S. and allied forces, DoD has embarked on a security program known as Navigation Warfare (NAVWAR). The three principal tenets of NAVWAR are to protect the use of GPS by DoD and allied forces in times of conflict within the theater of operations; prevent the use of GPS by adversary forces; and preserve routine GPS service to all outside the theater of operations.[32]

But it is in the nature of GPS that withdrawing one or more satellites from civilian use in order to deny their use to an 'enemy' must degrade – and in some circumstances preclude – civilian GPS access over a rather large area. This is especially true if some of the enemy uses envisaged would take place at high altitude, where more GPS satellites were in 'line of sight' than at ground level. It seems clear that the aim of preserving GPS access 'to all outside the theater of operations' can be accomplished only if the 'theater of operations' is drawn to extend far from the sites of actual combat.

Imagery and mapping

The US National Imagery and Mapping Agency anticipates exploiting access to commercial imagery. The other side of the question is whether US imagery and mapping can be available to the civilian sector. Of course, US mapping of the USA itself is publicly available, down to 1:24,000 topographic sheets, from the United States Geological Survey. What of maps created for military purposes? Apparently as a measure to ensure that Allied forces had ready access to material needed for 'battlefield visualization', NIMA has revised its security practices to permit expanded availability of 'national imagery at the unclassified level'.[33] Whether this availability goes beyond 'friendly' militaries is not clear.

Imagery denial and redundancy of sources

To the extent that verification of arms control and disarmament regimes comes to rely on satellite surveillance, states with an interest in those regimes gain by having access to reliable imagery. They could launch reconnaissance systems themselves, or do so jointly with others, or commission an inter-state body to do so, but only at great cost.[34] Non-state participants in disarmament practices also have a stake in access to trustworthy sources of surveillance data quite independent of those states whose activities they wish to monitor. Because of the cost of such systems, of course, states are reluctant to build and maintain them, and would prefer to rely on capabilities for which others have paid.

How serious is the concern that a monitoring government might deny salient imagery to other governments, or to public groups? The probability of denial is very high, and therefore the argument that there should be redundant sources – some of them in the hands of those most vigilant and committed to the arms control regimes – is very strong. In fact, on 22 July 1998 the USA, in what might be called 'pre-emptive denial', prohibited US firms from recording advanced satellite imagery of a US ally. The US Government told three US-based companies – Earthwatch Inc., Space Imaging Inc., and Orbimage – that they are banned from taking one-metre resolution images of Israel. Each had planned to orbit satellites with cameras capable of this higher degree of resolution. State Department officials explained this policy on 24 July as the consequence of Israeli fears that its 'enemies' would use the imagery. The companies may still take two-metre resolution images.[35]

While this is the first known constraint on 'public sector' imagery providers, it sets a precedent. It is almost certain that other states will now be prompted to ask for guarantees against surveillance. This need not be limited to allies. If the USA compels vendors to accept Israeli wishes, how will it explain to the UK or China that their sensitive sites should not be granted the same immunity? A further effect will be to spur development of one-metre resolution capabilities for launch by vendors not under US control.

CONCLUSION

The survey in this chapter supports the view that transparency is growing, and that elements of this transparency should make it easier to verify a Zero Nuclear Weapon regime. But the case is not simple or necessary, or certain in every respect, because sensing methods can be anticipated and countered, and because concealment can escape detection.

It is no exaggeration to say there is a far-reaching revolution in surveillance capabilities, roughly analogous to the revolution in military affairs. This has not come suddenly. Throughout the post-Second World War period, reconnaissance capabilities increased steadily, though the most advanced were confined to government control and (largely) military intelligence missions. John Lewis Gaddis remarked on this when he wrote of the 'reconnaissance revolution' as one reason for there being no major war between the powers since 1945.[36] This process continues to sharpen, first because of increasingly adept technology – applied across a range of sensing strategies – and second, because computing is exploited in novel ways to glean useful information from collected data. Two or more data sets taken at different time or frequencies, or by different methods altogether, may be exploited by using computational techniques to correlate the data sets and then discern meaningful difference.

To what extent does this revolution offer advocates of a nuclear disarmament, for example, greater possibilities to achieve an effective, persuasive regime? Are the Non-Aligned Movement, or Australia and Japan, in a stronger position than ten or twenty years ago to argue that a non-nuclear world suffers no greater risk of large-scale war than an ongoing nuclear world? It certainly seems so, but of course, that question suffers, as it must, from relying on a calculation of comparative future risk, something which may be prudently estimated but never truly known. Verification is also only one part of the fabric of political arrangements and positive assurances on which such a regime must fundamentally rely.

Whether an effective, persuasive disarmament regime can be achieved is also a larger question than whether there will be greater assurance that *militarily significant* departures from a Zero Nuclear Weapons regime, for example, are likely to be observed. On this more closely defined question, it does appear that a state seeking to build a clandestine nuclear force from scratch, or to render warheads (withheld from acknowledgment at an earlier date) suitable for performing military missions, would be more likely to be exposed at an early point in its activities if subject to new observing capabilities. Of course, much would turn on whether it was being examined closely. The most precise sensors will reveal nothing if there is not close, imaginative use of the information provided. But the tell-tale requirements of a manufacturing programme, and especially systems for prompt delivery of nuclear weapons, are unlikely to be on such a small and discreet scale that they would not prove detectable.

Sensing-processing systems offer growing capacities to disclose prohibited activity. This is not contradicted by the so-called 'US intelligence failure' at the time of the Indian nuclear tests of 1998 – a result of self-deception and resultant inattention. However, the Indian case does point out the need for watchers to realize that a state intending to flaunt a Zero Nuclear Weapons regime would tailor its activities to times and conditions in which discovery was more difficult.[37]

Of course, sensors and data interpretation practices need only point to possible sites, which must then be subject to direct inspection on the ground. Remote sensing and imaging offers only one among several ways in which forbidden, clandestine activities could be identified, and then must be followed by closer examination, and finally on-site inspection. A repertoire of many techniques, some of them not well understood by those subject to surveillance, is more likely to reveal prohibited activities, and even more to the point, is more likely to deter deception of a regime in place.

REFERENCES

1   Secretary of Defense, *Annual Report to the President and the Congress* (Washington, DC: US Government Printing Office, 1998), commonly referred to as the *1998 Annual Defense Report*. Citations here are to the on-line version: <http://www.dtic.mil/execsec/adr.98/index.html>. The URLs cited in the notes in this chapter identify sources consulted by the author. Some of these will no longer be current, but they have been included to indicate the basis for his assertions. Citations of printed versions have been included whenever possible.

2   *Scientific American*, Vol. 278, No. 4 (April 1998), p.23. The despatch identifies a Surface Heat Budget of the Arctic Ocean (SHEBA) project currently under way.

3   Don Belt, 'An Arctic breakthrough', *National Geographic* (February 1997), pp.36–57.

4   <http://www.odci.gov/cia/dst/html/peace_dividend.html>.

5   MEDEA Special Task Force Report, *Scientific Utility of Naval Environmental Data* (CD-ROM version available from Naval Meteorology and Oceanography Command, 1929 Balch Blvd, Stennis Space Center, MS, 39529–5005), 'Executive Summary'.

6   Ibid., para. II.D; 'Ocean Volume and Boundary Properties.'

7   Ibid., para. IIB: 'Geology and Geophysics.'
8   Jeffrey T. Richelson, 'Scientists in black', *Scientific American*, Vol. 278, No. 2 (February 1998), pp.48–55; p.55.
9   Ibid.
10 *New York Times*, 25 February 1995; *Boston Globe* service, 26 June 1995.
11 The Alace instruments, built by Webb Research Corporation, are deployed in the World Ocean Circulation Experiment (WOCE), part of the inter-governmental World Climate Research Programme. As of early 1998, 1,100 Alace floats had been built. WOCE is based at Southampton Oceanography Centre. Available at: <http://www.vsa.cape.com/~dwebb/alace.htm>.
12 <http://www.jpl.nasa.gov/releases/98/angkor98.html>. For other examples of the use of space-based photographs and radar images, see Farouk El-Baz, 'Space age archaeology', *Scientific American*, Vol. 277, No. 2 (August 1997), pp.60–65.
13 <http://www2.jpl.nasa.gov/files/images/browse/p49592bc.gif>.
14 The Defense Advanced Research Projects Agency (DARPA) and the US Air Force will jointly share the costs of developing and orbiting two synthetic aperture radar/ground moving target indicator satellites.
15 Associated Press, 10 March 1998, reporting an exhibit at the US Congress introduced by Secretary of Energy Federico Peña.
16 John Markoff, 'Using earth's magnetic pull to track ancient treasure', *New York Times*, 26 May 1998, p.B-9. Breiner was contacted by a scientist from Sandia National Laboratory. As Markoff tells it: '"Let's say we lost something," his caller said mysteriously. "Let's say we lost something at sea." What kind of metals can your magnetometers detect, the caller asked. It turned out that the Pentagon was frantically [trying] to find a hydrogen bomb that had accidentally fallen from a B-52 off the coast of Spain.' Sandia was aiding the quest for the weapon, but attributed the Navy's subsequent success to a nearby fisherman's observation and the use of manned submersibles. Available at: <http:www.sandia.gov/LabNews/LN01-19-96/palo.html>. Breiner's company is Geometrics: <http:www.geometrics.com>.
17 Secretary of Defense William S. Cohen, *Report of the Quadrennial Defense Review* (Washington, DC: US Government Printing Office, 19 May 1997); text available at: <http://www.defenselink.mil/pubs/qdr>. US Joint Chiefs of Staff (USJCS), *Joint Vision 2010* (Washington, DC: USGPS, 1997); text available at <http://usachcs-www.army.mil/DIV&INST/Jv2010.pdf>. *1998 Annual Defense Report*.
18 *1998 Annual Defense Report*, Chapter 7, 'Space Forces'.
19 Ibid.
20 Ibid., Chapter 8. This is spelled out in greater detail in Cohen, *Report of the Quadrennial Defense Review*, as follows: 'The five principal components of our evolving C4ISR architecture for 2010 and beyond are: (i) A robust multi-sensor information grid providing dominant awareness of the battlespace to our commanders and forces; (ii) Advanced battle-management capabilities that allow employment of our globally deployed forces faster and more flexibly than those of potential adversaries; (iii) An information operations capability able to penetrate, manipulate, or deny an adversary's battleship awareness or unimpeded use of his own forces; (iv) A joint communications grid with adequate capacity, resilience, and network-management capabilities to support the above capabilities as well as the range of communications requirements among commanders and forces; (v) An information defense system to protect our globally distributed communications and processing network from interference or exploitation by an adversary.'
21 *1998 Annual Defense Report*, Chapter 8.

22 Ibid.
23 Ibid.
24 USJCS, *Joint Vision 2010*, p.16.
25 *1998 Annual Defense Report*, Chapter 8.
26 *Independent*, 22 July 1998; Kevin Tebbit was named on 21 July 1998.
27 UK Ministry of Defence (UKMoD), Strategic Defence Review 1998, published as the White Paper, *Modern Forces for the Modern World* (London: HMSO, 1998). More detail is provided in Supporting Essays 3 and 5 in the supporting essay for this paper. Available at: <http://www.mod.uk/policy/sdr/index.htm>.
28 UKMoD, *Strategic Defence Review 1998*.
29 Ibid.
30 US Department of State, *Report on the International Control of Atomic Energy* (Washington, DC: 16 March 1956). Available at: <http://www.learnworld.com/ZNW/LWText.Acheson-Lilienthal.html>.
31 *1998 Annual Defense Report*, Chapter 7.
32 Ibid.
33 Ibid.
34 On 7 November 1998, Japan's Cabinet approved plans to launch four intelligence-gathering satellites by 2003. It had been alarmed by North Korea's missile launch across Japan, which revealed Tokyo's imperfect understanding of North Korean launch activities; *New York Times*, 7 November 1998.
35 Associated Press, 24 July 1998.
36 John Lewis Gaddis, 'The long peace: Elements of stability in the postwar international system', *International Security*, Vol. 10, No. 4 (Spring 1986), pp.99–142; 123ff. Of course, his observation centres on the USA and Soviet Union, with the result that lesser powers, some of whom fought fearsome wars, are not considered.
37 On the one hand, moving facilities underground seems to assure immunity from detection. Major portions of the Soviet nuclear weapon system operated underground. Concerns about North Korea's possible nuclear programme stem in part from the belief that North Korea has, or could have, underground facilities not known to the International Atomic Energy Agency. This argument is developed with respect to ballistic missile programmes by the Rumsfeld Commission, whose chair, former US Secretary of Defense Donald Rumsfeld, reported to a Congressional Panel on 28 July 1998 that North Korea, Iran, Russia, and China 'have made extensive use of the underground construction, which enables them to do things such as development and storage and, indeed, even launching from underground, hidden silo areas' (Reuters, 28 July 1998). Still, building and maintaining large-scale underground facilities does not guarantee they will go undetected. That must be the hidden subtext of the Rumsfeld report. If the sites were undetectable, the commission would have had nothing to say. And once revealed, either by observing traffic or material difference, or by disclosure, a site can be earmarked for ongoing scrutiny.

PART III
TECHNOLOGY TRANSFER FOR DEVELOPMENT

12 Technology Transfer for Developing Countries

Carlo Pietrobelli

INTRODUCTION: THE INCREASING ROLE OF TECHNOLOGY IN THE GLOBAL ECONOMY

Two major new features of social and economic systems have characterized the last two decades. On the one hand, technology increasingly plays a central role in all economic activities, with the rapid pace of technological change. On the other hand, all economic and technological activities have become global. The latter feature means that transfers and interchanges of technologies between countries have increased tremendously, and have become vital to their development.

These two dominant features are intrinsically interrelated and mutually reinforcing. Thus, the rapid pace of technological change induced by improvements in communications, transportation, information technology, and by new materials, is facilitating the international expansion of economic activities, which in turn further accelerates technological change. The product cycle of technology is getting shorter, and technology may be transmitted more quickly across countries. This can result in effective cycles of cumulative causation linking faster technological change to international competitiveness, to improved access to international knowledge, and to further technological changes. Such cycles may enable countries to catch up with technological leaders.

Technology has become a crucial production input, and the intensity of knowledge required for production has grown remarkably. Technological knowledge for production includes research and development (R&D), design, engineering and a host of organizational changes related to maintenance, management and marketing. Consequently, intangible investments – including R&D, training, software development, design and engineering – have been growing at three times the rate of tangible investments since the late 1970s.[1] New technologies such as microelectronics, biotechnology and new materials are creating new products,[2] while at the same time changing the characteristics and performance of many traditional products.[3] Moreover, the link and the causal relationship between technological change and competitiveness appears increasingly important.

The second dominant feature of the prevailing techno-economic system is the widespread internationalization of all economic and technological activities. For many years, foreign trade had been growing at a much faster rate than production. Technology is increasingly being exploited and generated globally, often as a result of international collaborations. Several economic parameters reflect this increasing internationalization of technology.[4]

Thus, the link between technological intensity and internationalization is becoming more important. Moreover, non-resident patents (patent applications by foreign inventors within a country, a proxy for the extent to which a country is 'invaded' by foreign inventions) and external patents (national inventors patenting abroad, a proxy for a country's inventions' 'invasion' of other countries) have grown at very high rates.[5] International exchanges of technological know-how and services have increased in relation to internal business R&D expenditure. The proportion of R&D carried out by transnational companies in their foreign subsidiaries is increasing.[6] International agreements today represent almost 60 per cent of all the registered inter-firm agreements.[7] This increasing internationalization is consistent with the nature of current technology, for which it is useful to extend the reach of a company's technological activities, to obtain technology abroad, and establish R&D and technology partnerships with other companies and institutions.[8]

This tendency towards globalization is reinforced by the 'generic' nature of many technologies. In fact, technologies often come from different disciplines and distant geographical areas; they are then combined in different ways for different uses. Hence, as the knowledge needed for efficient industrial production comes from many different sources, firms become less and less capable of independently supplying all the required technological knowledge. Therefore setting up all possible linkages in science and technology (S&T) and in R&D matter more than in the past.

In sum, the separation between the domestic and the international market for technology is being blurred, and firms need to be able to obtain their technologies globally. However, the enhancement of a country's technological and industrial development requires that the access of enterprises to international technology comes hand in hand with an effective strategy of technological and industrial development, supported by substantial local investments. Efforts to create and improve technological capabilities via investments in both general and technical education and training, and in S&T and R&D institutions, then become even more necessary.

Within this general context, the aim of this chapter is to look at the notion of international technology transfer and its main channels of transmission, and to examine some of the overall strategies related to it, along with their implications for the overall industrialization strategy of developing countries. Some data is presented to illustrate the alternative options that are being followed by selected developing countries.

The chapter offers a brief review of the main transmission channels of foreign technology, with some possible categorizations of them, then makes explicit the underlying concept of technology and technological change that is employed when observing and analysing technology transfer. This helps to clarify the policy implications of the various strategies of foreign technology acquisition. Some criteria for the selection of the appropriate mode of transfer are singled out and discussed to test how well these theoretical considerations match the empirical evidence. We then present some preliminary evidence on a sample of countries that illustrates some of the many possible strategies of foreign technology acquisition and local industrial and technological development. The final section summarizes the evidence and draws some general policy conclusions.

TRANSMISSION CHANNELS OF FOREIGN TECHNOLOGY

Technologies make their way across international boundaries in a variety of ways, ranging from technologies embodied in capital equipment and learned from knowledgeable buyers to inter-firm agreements and foreign investments. Globalization of technology implies that this variety is itself still growing, and that technology may be transferred through an even larger number of different mechanisms.

Several categorizations of transfers have been proposed, each focusing on one or more dimensions of each channel. Some of these categories may overlap, and the definitions are not mutually exclusive and have been used differently by various authors. In addition, business operations involving technology flows usually incorporate several channels of international technology transfer at once. However, setting out the definitions of these categories should contribute to a better understanding of this complex phenomenon.

Informal versus formal channels

The acquisition of foreign technology from abroad can be formal or informal, depending on whether it is paid for and subject to a contract, or simply transferred through observation, publications, imitation or embodied in skilled people shifting from one country to the other. Foreign buyers often represent an effective channel through which technologies flow informally (whether on purpose or not) to manufacturers in developing countries.[9] This chapter focuses only on *formal* channels of technology transfer.

Formal co-operative agreements versus arm's-length transfers (internalized versus externalized methods)

The latter include arm's-length trade in machinery, components, intermediate inputs and direct purchases of knowledge (such as patents or blueprints), whereas the former comprise formal co-operation between a domestic and a foreign firm. Other authors instead distinguish between internalized and externalized means of technology transfer.[10] However, not all technologies are available at arm's length, and due to their inherent complexity or strategic value, may only be obtained through internalized means or direct formal co-operation – through majority ownership, joint ventures and other forms of strategic alliances.[11] Among technology transfer channels involving interfirm relationships, an additional useful distinction may refer to the equity or non-equity nature of the agreement – whether it involves ownership participation or not. Substantial evidence suggests that the latter kind has considerably expanded in recent years.[12]

Embodied versus disembodied

Technology transfer channels differ depending on whether or not the technology is embodied in physical equipment, requiring differing amounts of human capital to 'disembody' them into technical knowledge useful for efficient production. Most frequently, international flows of technology transfer include both forms, because disembodied technology is necessary to use embodied technology efficiently (for example, technical assistance must flow to developing countries together with imported capital equipment). This distinction is related to the following one of packaged versus unpackaged technology.

Packaged versus unpackaged

This distinction derives from a multi-dimensional view of technology that goes beyond the simple notion of production technique: that is, the combination of productive inputs that go into the production function. Technology is itself made up of several elements that in turn affect its efficient deployment. It reflects the social environment where it is introduced and used, as well as the physical capital, qualified personnel and all the necessary technical and organizational knowledge.

Packaged and unpackaged international technology transfers differ depending on whether the transferred technology arrives along with all the elements required to ensure its effective application. These elements range from physical capital to organizational and managerial knowledge. As a result, the requirements of local technological efforts and capabilities differ remarkably in each case.

This distinction between packaged and unpackaged technology transfer proves especially useful when exploring the role played by international technology flows in a country's industrial development strategy. In this context, the major forms of formal technology transfer from abroad are: international trade (imports of capital goods), foreign direct investment (FDI) and majority joint ventures, and foreign licensing, together with the broad range of international inter-firm and inter-institutional agreements. They vary in the extent of packaging – whether they arrive together with all the elements required to ensure the effective use of the technology.

Thus, under FDI, technology arrives as a package, including the capital equipment and the foreign company's managers and engineers. Little else but cheap labour and some intermediate technical skills are required to use such transferred technology efficiently for simple industries. In this case, technology is transferred through *internalized* means, and the least-developed local capabilities are required. Therefore, FDI is the easiest way in which to import foreign technology, but it occurs at the risk of not transferring capabilities. These are likely to remain within the originating transnational company, which would result in excessive and permanent dependence on the foreign company.

The import of capital equipment alone is the extreme opposite of unpackaging, involving the transfer of only some physical elements of the technology package, and relies fully on local engineers and technicians, and on local S&T institutions for its efficient application. The resulting costs of lengthening the period of inefficient deployment of the imported capital goods may be compensated by the permanent learning acquired by local experts.

Minority joint ventures, other forms of strategic/technology partnerships, management and marketing contracts, international subcontracting, foreign licensing and technical assistance agreements are intermediate options between the extreme alternatives of FDI or unpackaged international trade.

On the issue of technology transfer through FDI, several authors distinguish usefully between different types of R&D activities carried out abroad by transnational companies.[13] They identify four main different types of foreign R&D units of transnational companies:

1 technology transfer units (TTUs) that facilitate the transfer of the technology of the corporate parent to a subsidiary, and provide local technical services
2 indigenous technology units (ITUs) that develop new products for the local market, drawing on local technology
3 global technology units (GTUs) that develop new products and processes for major world markets
4 corporate technology units (CTUs) that generate basic technology of a long-term or exploratory nature for use by the corporate parent.

These types can be classified in terms of their increasing ties with local S&T institutions. The potential for the diffusion of new knowledge to the local S&T system is greater through GTUs and CTUs. However, the other types of R&D units are important for transferring foreign technology back to the host country. Recent evidence on India suggests that the scope of R&D activities by transnational companies within that country has broadened, due to the locally available supply of educated high-tech personnel at costs much lower than at other international locations.[14]

What is the relative quantitative importance of these possible channels of technology transfer? Data for 1990 reveal that the importation of capital goods represented over three-quarters of the total international flow of technology transfer to developing countries (see Table 12.1). FDI flows were

Table 12.1   Different modes of technology transfer to less-developed countries in 1990

			Growth rates	
	US$billion	*%*	*1980–84*	*1985–90*
FDI inflows	32	15.8	3	19
Royalties and fees	2	1.0	5	12
Capital goods imports	155	76.7	–2	11
Technical co-operation	13	6.5	1	12
Total	202	100.0		

Source: UN World Investment Report (New York: UN, 1992).

only 16 per cent of the total flow to less-developed countries, in spite of their faster growth rate of 19 per cent in the period 1986–90. Comparable data for the 1990s is not yet available, but we would expect to find evidence of growing rates of all modes of technology transfer, but especially of foreign direct investment. The international expansion of enterprises' activities has represented a major feature of the prevailing techno-economic paradigm.[15]

THE CONCEPT OF TECHNOLOGY UNDERLYING TECHNOLOGY TRANSFER

The technology market has often been studied by economists with the tools that are used to analyse a competitive market. If technology might be studied like any other commodity, and if markets worked freely and perfect competition prevailed, then the analysis of technology transfer would pose no problems. Technology (from whatever source) would be easy to transfer and use. The efficiency of its use would only be a matter of ensuring the conditions for efficient resource allocation in the context of externally determined technological alternatives. Technology policy would then consist only of government sponsorship of institutes that collect, process and disseminate technical information: activities justified as a provision of public goods. This conception derives from two assumptions: that technology consists simply of a set of techniques wholly described by their 'blueprint', and that all techniques are created in the developed countries, from which they flow to developing countries.[16]

However, several authors recognized early on the special features of technology and technological change, and arrived at a more complex perception of technology. First, no existing technique is completely expressed by the sum and combination of its material inputs and the codified information about it. In fact, much of the knowledge about how to perform elementary processes and about how to combine them efficiently is *tacit*, not feasibly embodied, nor codifiable nor readily transferable, and 'a firm will not be able to know with certainty all the things it can do, and certainly will not be able to articulate explicitly how it does what it does'.[17]

This means that technology is not simply a set of blueprints or instructions which, if followed precisely, will always produce the same outcome. Although two producers in the same circumstances may use identical material inputs with equal information available, they may none the less employ two very distinct techniques due to their different understanding of the tacit elements. Techniques are sensitive to specific physical as well as social circumstances.[18] Moreover, technology is not instantaneously and costlessly accessible to any firm: a firm does not simply select the preferred option from the freely available international technology, as there may be obstacles and difficulties in obtaining the desired technology.

Furthermore, simply choosing and acquiring a technique does not imply the ability to operate it efficiently (at the level of 'best practice'). Individual firms do not have a complete knowledge of all the possible technological alternatives, their implications, and the skill and information they require. The entire production curve, illustrating an infinite number of alternatives, is *not* known to the individual firm, as neo-classical theory assumes. To the extent that technologies are tacit, firm production sets are 'fuzzy' around the edges.[19]

Understanding technology in these more complex and realistic terms implies that investments in technology are required whenever technology is newly applied. This applies to domestic as well as foreign imported technologies. Each firm has to exert considerable absorptive efforts to learn the tacit elements of technology, and gain adequate mastery. This is at the opposite extreme from the neo-classical premise that technology, as well as productive inputs and outputs, is perfectly known. This knowledge is not instantaneously and costlessly available to all firms, and technology transfer poses substantial problems of adaptation and absorption that are related to investments in technological capability – to the complex array of skills, technological knowledge and organizational structures required to operate a technology efficiently and accomplish any process of technological change.[20] This dynamic technological effort implies a process of learning that is qualitatively different from the traditional 'learning by doing', as it involves an active attitude. Learning may be pursued in a variety of ways,[21] and the passive 'learning from operating' is only one of many possibilities.[22]

In addition, even if the need for learning efforts is acknowledged, investing in learning does not ensure success. This is due to the stochastic nature of the learning process, which is influenced by the external environment and by a firm's actions, and results from dependence on historical circumstances, entrepreneurial skills and luck. Therefore, different firms may reach persistently different levels of efficiency and dynamism, even in competitive markets.[23]

Within this broader context, technology transfer becomes an important issue that has to be assessed jointly with a country's capability to make use of technology, absorb it and adapt it to local conditions. In other words, technology transfer links foreign technology access and acquisition to its efficient use for economic development, and to the effort of the relatively technologically backward countries to catch up.[24]

Technological efforts

Clearly, the access to and acquisition of foreign advanced technology is not in itself sufficient to ensure local technological and industrial development; several other elements are needed. An additional vital component of a country's industrial development policy strategy is the technological effort oriented to the absorption, adaptation, mastery and improvement of technology. This implies the need for a continuous process of technological change.[25]

When measuring technological effort, one should take into account all the efforts of assimilating technology, adapting it to local conditions, mastering production processes and designs, trouble-shooting at the shop floor level, and experimenting with new products and processes. Unfortunately, only very aggregate measures of these efforts are available, such as the expenditure on formal R&D, or the local personnel employed in R&D. These are more important at higher levels of industrialization with high-tech, large-scale industries. In these industries, the role that R&D plays in developing a firm's learning and absorptive capacity – its ability to identify, assimilate and exploit knowledge from the environment, and not only to create new knowledge – must not be underestimated.[26] Measures of R&D are essentially input indicators. In contrast, indicators of output of successful technological efforts range from patents to new products and new exports.[27]

A precondition demanded by business enterprises to persuade them to exert the necessary technological efforts is the availability (and the ongoing creation by the local educational system) of human general and technical skills. Proxies for this availability are the enrolment rates at various educational levels, whose importance rises with the sophistication of local industry. In addition, enrolment rates should be looked at together with indicators of the quality and effectiveness of education, such as the pupil/teacher ratio and the percentage of pupils attaining advanced grades.

Considering the complex nature of technology described above, any national strategy to import technology from abroad is intrinsically related to local technological efforts and skills creation. Such efforts can be measured by some data, as shown in the next section.

Criteria for the selection of the channels for technology transfer

How should we choose between the different channels of technology transfer? Choice implies that there are different skills available, and that different policy strategies are related to the acquisition of these skills. Thus, technology transfer has to be viewed in the context of a country's overall industrial development strategy. However, choice is also conditioned by some specific factors geared to the organization of industry and the characteristics of the agents involved.

Lall has presented some considerations a country might take into account in choosing a preferred mode of technology transfer.[28] The following four considerations are among the main factors affecting the best choice of a channel of technology transfer.

The nature of technology

This may be described in terms of the following characteristics:

1   its inherent complexity, and the need for continuous interaction with the technology transferrer
2   the rate at which it changes – how frequently and rapidly the technology is likely to change
3   the newness of the technology
4   the degree of centralization of the required R&D – to what extent the nature of technology makes it preferable to carry out the required R&D in a centralized location, such as at the TNC headquarters
5   whether the technology relates mainly to the product or to the production process.

Considering these characteristics of technology, we would expect that technology transfer will take a more internalized form the more complex it is, the faster and more frequently it changes, the newer and more innovative it is, and the greater the convenience of carrying it out in a centralized location and in a large-scale facility. In very general terms, process technologies are often externalized, as the latest version of a technology lies with engineering firms. In contrast, product technologies are often internalized, as they tend to be more demanding in terms of skills. However, exceptions to this general pattern are frequent.

The strategy of the seller

This also inevitably influences the enterprise's selection of the technology transfer channel. If the seller firm is very large, the best technology transfer strategy may be internalized, as the seller has more capacity to handle transnational operations. If the seller's products are more diversified, then the technology transfer strategy might be more externalized, focusing on non-core technologies. If the seller possesses proprietary brand names of high value, then the technology transfer strategy might be more internalized. If the seller has less experience with technology transfer, the technology transfer strategy might be more externalized. If the seller's strategy is affected by the home corporate culture, the technology transfer strategy would have to be selected on a case-by-case basis, depending on this culture.

The capabilities of the technology buyer

Given the nature of technology as a package, these capabilities will also be reflected in the choice of the technology transfer channel. Thus, a more capable buyer will impose lower efforts and costs on the seller in an effective transfer of technology and the related knowledge. This would require less internalization, but would also pose a greater potential threat to the seller, especially to one who is export-oriented and possibly competing for the same markets. In this case, internalization may be preferred, after tight bargaining, otherwise the sale would be likely to involve very high prices and many limiting conditions.

Host government policy

This may also affect the choice of the channel of technology transfer in important ways. Traditionally, the tendency to prefer externalized technology transfer in order to protect the development of local capabilities (for example, via import tariffs, laws controlling and regulating or even impeding foreign direct investment) has prevailed. Since the 1980s, with liberalization and export-oriented economic strategies prevailing in most countries, more liberal policies towards trade and FDI have been increasingly adopted.

No clear-cut recipe can be given for the choice of the appropriate technology transfer channel. Learning benefits and externalities may induce one to prefer externalization. However, this applies only in so far as the host country already possesses some elements of the technology package, and if it is simultaneously investing in such skills and institutions, and has already been doing so for some time. In the history of developing countries, many different strategies have been followed, with widely differing results.

DATA ON TECHNOLOGY TRANSFER STRATEGIES AND INDUSTRIAL AND TECHNOLOGICAL DEVELOPMENT

A sample comprising several countries from different regions has been selected to illustrate the variety of strategies actually followed by developing countries. The selection aims to give an overview of interesting and differing stories of technology acquisition and local industrial development.

The following groups of countries are examined:

1   lesser-developed sub-Saharan African (SSA) countries, including a large
    and poor country that has been relatively more closed to foreign eco-
    nomic relations for many years (Ethiopia), a country that experienced
    early structural adjustment with substantial foreign assistance, often
    quoted as a success story of World Bank-supported adjustment policies
    (Ghana), and three countries of Southern Africa – the emerging power in
    Africa (South Africa), one of the most industrialized in the continent
    (Zimbabwe), and a new and promising country (Namibia)
2   a large South Asian country that followed inward-oriented policies for
    many years, and that is now pursuing an open-door (India)
3   a group of Asian newly industrialized countries (NICs), including some
    that are old (South Korea, Singapore and Taiwan) and some that are
    newly emerging but rich in resources (Malaysia and Thailand)
4   a group of Latin American countries, including two large (Argentina and
    Brazil), and two smaller (Chile and Colombia) economies, all with long-
    established industrial sectors
5   Hungary, as an isolated example of a country in transition to a market
    economy.

Some data for Italy and the USA are presented as useful comparators with
the situation in industrial countries.

The basic indicators for the sample countries vary enormously, both
among the different groups, and within them. Thus, Ethiopia recorded the
lowest GNP per capita (US$100 at 1995 rates), and Singapore the highest
(US$26,730), although estimates using the purchasing-power-parity method
partly reduce this gap. Growth rates also differ remarkably, as well as
economic sizes (see Table 12.5).

For each of these countries, several statistics relevant to technology transfer are displayed. These statistics may be divided into three sections: the
first relates to the technology transfer channels preferred by these countries;
the second relates to the technological efforts exerted by each to absorb,
adapt and improve technology and to the level and quality of investments in
education and human capital, and the third relates to the achieved economic
and manufacturing performance.

To examine the blend of technology transfer channels preferred by these
countries, data is supplied on FDI inflows as a percentage of GDP and of
gross domestic capital formation, on strategic alliances involving developing countries, and on imports of capital equipment.

The role of FDI

Many developing countries rely substantially on the most packaged form of technology transfer – FDI (see Table 12.2). In Sub-Saharan Africa, this is the case with Ghana and Namibia, where foreign capital is responsible for a notable share of domestic investments.[29] Ethiopia has never been a major recipient of FDI. In Asia, Singapore and Malaysia are the largest recipients. In stark contrast, South Korea and Taiwan have traditionally been rather impermeable to foreign investors, which have been allowed to invest in the country only under tightly controlled conditions.[30] For many years, India has been following a policy of protection from transnational companies, which have been allowed to operate in the country only in joint ventures with local companies, and have been subject to stringent controls. Among our sample countries in Latin America, Chile has been the most open to FDI.[31] Brazil has been relatively independent of foreign capital as a fraction of its largest economy, in spite of the large absolute values of FDI inflows.

Scattered evidence from several sources reveal that *inter-firm strategic alliances* are emerging as a remarkable channel of international technology transfer, also in the case of developing countries.[32] However, the empirical evidence available on this phenomenon is rather scant, old and unsystematic. Possibly the only exception to this is the Co-operative Agreements and Technology Indicators (CATI) database, constructed at MERIT, University of Maastricht, on the basis of the economic press in Europe and the USA since 1980.[33] Studies based on this information reveal that most developing countries are not participating in this new trend: enterprises from LDCs are involved in only 1.5 per cent of the 4,192 strategic alliances recorded in the database, and enterprises from NICs participate in only 2.3 per cent of the total.[34] Firms from the NICs appear rather active in sectors such as the automotive, food and beverages industries, and information technology, but much less so in biotechnology, software and aviation. Technology transfer to developing countries based on inter-firm agreements is therefore confined to very few more industrialized countries, such as South Korea, Taiwan, Singapore and Brazil.

Imports of capital goods

The data on imports of capital goods, while incomplete, sketches an interesting pattern (see Table 12.3). The Asian newly-industrialized countries (NICs), as well as Chile, appear to rely on unpackaged technology transfer, as the data on South Korea and Thailand suggest. Payments for royalties and licence fees reinforce this conclusion (but not for Chile). In the cases of

Table 12.2   Foreign direct investments during 1980–95 for selected countries

Ranking	Country	FDI inward stock as % of GDP				FDI inflow as % of GDCF	
		1980	*1985*	*1990*	*1995*	*1985–90*	*1995*
LDCs excluding China		4.8	9.0	9.2	15.0	6.7	5.9
Sub-Saharan Africa							
2	Ethiopia	2.7	2.4	2.0	2.4	0.3	1.1
31	Ghana	1.5	4.3	5.1	15.8	17.8	22.2
38	Zimbabwe	n.a.	n.a.	–0.9	1.1	1.8	3.1
76	Namibia	n.a.	1.3	2.3	12.3	22.9	6.7
91	South Africa	20.4	19.1	7.9	7.8	0.7	0.1
27	India	0.7	0.5	0.5	1.9	1.2	3.6
Asian NICs							
83	Thailand	3.0	5.1	9.3	10.3	10.2	2.9
99	Malaysia	24.8	27.2	33.0	52.1	43.7	17.9
108	Korea Rep.	1.8	1.9	2.3	2.3	1.9	1.1
	Taiwan	5.8	4.7	6.2	7.3	5.1	2.7
126	Singapore	52.9	73.6	76.3	67.4	59.3	24.6
Latin America							
75	Colombia	3.2	6.4	8.7	12.1	17.0	14.8
96	Brazil	6.9	11.3	8.1	17.8	3.1	4.7
101	Chile	3.2	14.1	33.1	23.1	21.5	10.8
105	Argentina	6.9	7.4	6.2	8.7	13.0	11.7
100	Hungary	n.a.	n.a.	6.3	31.5	33.3	59.7
Developed Countries		4.8	6.0	8.3	9.1	5.5	4.4
118	Italy	2.0	4.5	5.3	5.7	2.6	2.1
128	USA	3.1	4.6	7.2	7.7	5.3	5.9
World		4.6	6.4	8.3	10.1	5.4	5.2

Notes: Ranking is by per capita GDP according to the World Bank; GDCF is Gross Domestic Capital Formation.

Source: UNCTAD, *World Investment Report 1997* (Geneva: UNCTAD, United Nations, 1997).

Table 12.3   Imports of capital goods as a channel of technology transfer

	As % of GDP				As % of GDI		As % of MVA	
	1970	*1980*	*1990*	*1994*	*1965*	*1986*	*1965*	*1986*
Mauritius	4.8	8.9	20.2	15.9				
Kenya	8.3	10.2	10.2	9.0				
India	0.7	1.0	1.5	2.2				
Thailand	6.6	7.4	16.7	17.9	29.0	28.6	35.4	30.5
Malaysia	11.6	17.8	36.0	42.2				
Korea Rep.	6.5	8.4	10.4	10.8	16.1	37.4	10.4	36.6
Taiwan	8.9	10.3	13.8	14.8				
Singapore	32.7	66.2	84.4	87.2				
Mexico	3.5	4.5	7.9	16.8				
Chile					22.6	43.0	13.6	19.3
Brazil	2.5	2.2	2.2	1.9				
Colombia					20.1	20.2	22.2	22.4

Notes: GDI = gross domestic investments, MVA - manufacturing value added.

Source: Database in UNIDO, *Industrial Development Global Report 1996* (Vienna: UNIDO, United Nations, 1996). C. Pietrobelli, *Industry, Competitiveness and Technological Capabilities in Chile: A New Tiger from Latin America?* (London and New York: Macmillan and St Martin's Press, 1998), table 4.9.

Malaysia, Singapore and Mexico, the remarkable levels of imports of capital equipment are perhaps related to the many transnational companies operating in these countries. In contrast, machinery imports are only very small shares of GDP in Brazil and India – countries that have larger domestic manufacturing sectors. However, the former pays large amounts in royalties and licence fees.

Technological efforts

The data on the technological efforts exerted to absorb, adapt and improve technology, and on the level and quality of investments in education and human capital, is equally diverse from country to country (see Table 12.4). Yet a broad general pattern emerges, with countries in Sub-Saharan Africa exerting minimal technological efforts, much less so than the advanced newly industrialized countries and also Latin America. Within the Asian NICs, South Korea and Taiwan appear to devote massive resources to technology, much more than Malaysia and Thailand, which appear more open to internalized forms of technology transfer. To a lesser extent, Hungary also appears to invest resources in technology development, and it has fairly major patenting activities, probably a heritage of the Soviet-style planning system.

Education

Investments in education by the African countries are not yet sufficient to allow them to catch up with the technological leaders (see Table 12.4). It is worrying that this conclusion is reinforced by the data on the poorer quality and effectiveness of education in these countries. The levels of tertiary education attained in Korea and Taiwan are remarkable, as they are higher than in Italy and on a par with Argentina, a country that has had a system of free universities for nearly a century. Other data show that the scientific orientation of higher education is much more pronounced in these countries.[35]

In principle, effective access to foreign technology needs to be accompanied by local investments in technology and education. These investments are absolutely necessary if technology is imported through externalized means (such as capital equipment imports, purchase of patents and licenses, and inter-firm agreements where local control is retained), since some additional elements of the technology package must be provided locally. Furthermore, these investments are also a necessity if the country wants to learn effectively from foreign investors, and wants to transfer industrial capabilities from abroad permanently.

Economic and industrial performance

As a consequence, it is not possible to make clear-cut judgements about the particular policies adopted for technology transfer, the development of technology and education and the observed economic and industrial perform-ance achieved. However, it is instructive to examine some indicators of performance for our sample countries (see Table 12.5). The growth rates of GDP and manufacturing in general appear higher for the Asian NICs, and Chile and India. The Asian NICs (and especially those which emerged in the 1990s) have consistently recorded higher growth rates of exports than most other countries in our sample. Moreover, their export structure has been evolving more rapidly towards manufactured exports, the most dynamic item in international trade during the last decade.

The scattered (and imperfect) data on manufacturing value added (MVA) and productivity are consistent (see Table 12.5), in that they also suggest that both MVA and labour productivity in manufacturing have evolved more rapidly in East Asia than in Latin America, and that South Africa does not appear to be very far behind.[36]

SUMMARY AND CONCLUSION

We can now summarize a taxonomy of strategies implemented by some developing countries. The taxonomy is developed along two main dimen-sions: one is the prevailing choice of packaged or unpackaged modes of technology transfer, the other is the strong or weak local efforts in technol-ogy development and improving the educational system. Countries have been classified along these two dimensions, with some details given in each case.

Strong local technological effort

The countries have mostly packaged channels of technology transfer:

- **Malaysia**: investments in technology/communications infrastructure, advanced technical education to attract foreign direct investment (FDI)
- **Singapore**: similar to Malaysia, more involved in high technology

These countries have mostly unpackaged channels of technology transfer

- **Korea**: comprehensive industrial and technology policies, large con-glomerates, bargaining with foreign investors
- **Taiwan**: comprehensive industrial and technology policies, often via State-owned institutes, mainly Small and Medium-sized Enterprises, more open to FDI.

Table 12.4 The technological and educational base, 1980–95, for a selected sample of countries

	R&D		Royalty & licence fees US$million		Patent applications Residents		Education (% of cohort)		
	S&E per million 1981–95	R&D as % of GNP 1981–95	Receipts 1996	Payments 1996	Yes 1995	No 1995	Secondary 1992	Tertiary 1993	Reach Grade 4 1990–1
Sub-Saharan Africa									
Ethiopia			0	0			11	1	28
Ghana	349[1]	0.9[1]	1	1		42	36[6]	1.8[3]	
Zimbabwe			0	6		177	47	6	77
Namibia				5			55	3	64
South Africa			67	250	5,549	5,501	74	13	
India	151	0.8	2	90	1,545	5,021	48	5[1]	
Asian NICs									
Thailand	173	0.2	25	717	62[4]	n.a.	37	19	85
Malaysia	87	0.4	0	0	141	3,911	60	7	100
Korea Republic	2,636	2.8	185	2,431	59,249	37,308	91	48	100
Taiwan	1,426	1.8[7]						37	
Singapore	2,512	1.1			10	11,871	79	14	100
Latin America									
Colombia	40[2]	0.1[7]	59	49	141	1,093	61	10	74
Brazil	165	0.4	32	529	2,757	23,040	43	12	
Chile		0.8	63	51	181	1,535	69	27	95
Argentina		0.3	6	221			81[5]	41	98

Hungary	1,157	1.0		45	132	1,117	19,770	79[5]	17
Developed Countries									
Italy	1,303	1.3	381	1,027	1,625	63,330	81	37	100
USA	3,732	2.9[5]	29,973	7,322	127,476	107,964	97[5]	81	100

Notes:
1   1980
2   1982
3   1985
4   1986
5   1988
6   1991
7   1993

S&E = scientists and engineers.

Sources: The World Bank, UNESCO and WIPO.

Table 12.5 Growth rates of GDP, manufacturing and exports, manufacturing value added and labour productivity for selected countries, 1980–96

	Average annual growth			Manufactured exports	MVA	Labour productivity
	GDP 1990–6	Manufacturing 1990–6	Exports 1990–5	% of total exports 1993	1995[1]	1995[1]
LDCs – China & India	7.6	13.2	2.7			
Sub-Saharan Africa						
Ethiopia	3.9	3.3	−9.4	3.7		
Ghana	4.4	2.6	9.1	23.4		
Zimbabwe	1.3	−3.7	−6.6	36.5	116.6	106.0
Namibia	4.1	3.8				
South Africa	1.2	0.8	2.8	73.5	107.6	158.2
India	5.8	7.5	7.0	75.1		
Asian NICs						
Thailand	8.3	10.7	21.6	72.6	430.3	364.6
Malaysia	8.7	13.2	17.8	64.8	450.1	229.0
Korea Republic	7.3	7.9	7.4	93.2	454.9	671.3
Taiwan			5.9	93.1	252.2	472.9
Singapore	8.7	7.9	16.2	80.0		

Latin America						
Colombia	4.5	1.4	4.8	39.8	160.8	157.6
Brazil	2.9	2.2	6.6	59.5		
Chile	7.2	6.3	10.5	18.3	171.3	192.3
Argentina	4.9	5.0[1]	−1.0	31.9		
Hungary	−0.4	1.1[1]	−1.8	68.0		
Developed Countries						
Italy	1.0	1.0	6.0	89.1		
USA	2.4	1.6	5.6	81.6		
World	2.2	1.4	6.0			

Note: 1 1980 index is 100.

Sources: World Bank, *World Development Indicators 1998* (Washington DC: The World Bank, 1998) and various annual reports of UNIDO world industry development indicators.

- **India until opening**: closed for many years to foreign capital, technological and education efforts in State-owned institutions, never facing the incentive to compete in international markets.

These countries have both packaged and unpackaged technology transfer, depending on what is available on the international technology market:

- **Chile after the early 1990s**: increase in FDI inflows and outflows, technological efforts are increasing with some government interventions
- **Hungary**: very open to FDI and alliances, also to imports with special arrangements with the European Union, large, but often old and inadequate technology structure, good general education
- **Brazil**: State-oriented local technological efforts, large local State-owned R&D centres, large local monopolies under import protection, little exposed to international technology flows in whatever form; now gradual opening, privatizations, foreign investors, cheaper imports, with continuing public technology support system.

Weak local technological effort

These countries have mostly packaged technology transfer:

- **Thailand**: very open to FDI – often of assembly type, shortage of intermediate technical skills, still little support for local technological development; recently efforts have increased
- **Argentina until *c*. 1990 (during import-substituting industrialization and hyper-inflation)**: high import tariffs, more open to FDI, hardly any support to local technological development, good human capital base but insufficient scientific orientation – with less technological efforts than Brazil.

These countries have mostly unpackaged technology transfer:

- **Chile until the early 1990s**: few FDI, limited technological efforts, good base of human capital but little scientific orientation, ideological neglect of need for technology support.

These countries have both packaged and unpackaged technology transfer, depending on what is available on the international technology market:

- **Sub-Saharan Africa – South Africa, Ghana, Namibia, Zimbabwe**: open to international markets – Ghana to a greater extent, but very little technology support, insufficient human capital for advanced industry
- **Ethiopia**: still very closed to international technology flows, and insufficient investments in technology and education
- **Argentina after *Convertibilidad* (1991)**: more open to international exchanges and to FDI, fixed exchange rate, makes imports cheaper, still little attention paid to technology development.

In general, for a country with interests in technology transfer from the outside, there are two broad options for upgrading technology and industrial development. One option is relatively *dependent*, based on internalized technology transfer, essentially via foreign direct investment. The other option is to be relatively *autonomous*, based on indigenous enterprises with externalized technology transfer.[37] The latter is more demanding in terms of local capabilities, technological efforts, and policy and institutional support, but is more likely to promote local technological and industrial development. Countries following such a strategy include South Korea and Taiwan.

A strategy relying on foreign direct investment as the main channel of technology transfer can lead to remarkable technological improvements and competitiveness of local industry, but only if the government selectively stimulates these improvements of activities by negotiating with foreign investors, and supporting them through the creation of local human capital and technology infrastructures. Remarkable examples of this strategy are Malaysia and Singapore; Hungary also appears, to a lesser extent, to follow a similar route.

However, such a strategy may result in a technologically 'shallow' local industry that is potentially vulnerable to autonomous decisions by transnational companies, and hence an intrinsically fragile industry. This appears to be the case with Hong Kong, and could be the case with other countries such as Argentina, Ghana or Thailand if their governments and private sector do not devote substantial resources to local technological capabilities and infrastructures. Chile appears to be vacillating between the two latter options. Sub-Saharan Africa still requires substantial investments in technology, education, and wider access to foreign advanced technology. The latter appears also to be necessary for large countries like Brazil and India, which in the past often relied on domestic technology.

As is clear from the data above, the sample countries have had different degrees of success in their industrialization and economic development. No strategy guarantees success, but the coherence of the overall strategy is crucial. Thus, any strategy must take into account in a realistic way the possible external constraints in the access to international technology, as well as the requirements of each strategy in terms of local investments in technology and education.

It is apparent that technology is increasingly important in economic activities, and that the technology market has become internationally integrated and global. Developing countries need to be able to obtain their technology in this international market. However, access to foreign advanced technology is not sufficient to ensure rapid industrial and technological development. One conclusion is clear: for successful technology transfer, access to foreign technology must be accompanied by substantial local efforts to create and improve technological capabilities and good human capital.

REFERENCES

1 OECD, *Technology and the Economy: The Key Relationship* (Paris: OECD, 1992).

2 United Nations, *World Economic and Social Survey 1995* (New York: United Nations, 1995).

3 UNCTAD, 'New technologies and issues in technology capacity building for enterprise', (Geneva: UNCTAD, Division for Science and Technology, mimeograph, 1995).

4 D. Archibugi and S. Iammarino, 'Innovation and globalisation: Evidence and implications', mimeograph on the EU-TSER Project on Technology, Economic Integration and Social Cohesion. D. Archibugi and J. Michie (eds), *Technology, Globalisation and Economic Performance* (Cambridge: Cambridge University Press, 1997).

5 Archibugi and Iammarino, *Innovation and Globalisation.*

6 P. Reddy, 'New trends in globalization of corporate R&S&D and implications for innovation capability in host countries: A survey from India', *World Development*, Vol. 25, No. 11 (1997), pp.1,821–37.

7 C. Freeman and J. Hagedoorn, 'Catching up or falling behind: Patterns in international interfirm technology partnering', *World Development*, Vol. 22, No. 5 (1994), pp.771–80.

8 L.K. Mytelka (ed.), *Strategic Partnerships and the World Economy* (London: Pinter Publishers, 1991). C. Pietrobelli, *Emerging Forms of Technological Co-operation: The Case for Technology Partnerships – Inner Logic, Examples and Enabling Environment,* Science and Technology Issues (Geneva: UNCTAD, United Nations, 1996).

9 See C. Pietrobelli, *Tecnologia e Sviluppo: L'inserimento internazionale di un'economia emergente* ('Technology and Development: International Inclusion of an Emerging Economy') (Rome: Edizioni Lavoro, 1991), Chapter 4. D. Keesing and S. Lall, 'Marketing manufactured exports from developing countries: Learning sequences and public support', in G.K. Helleiner (ed.), *Trade Policy, Industrialisation and Development* (Oxford: Oxford University Press, 1992).

10 S. Lall, 'The interrelationship between investment flows and technology transfer', Issues Paper for the UNCTAD Working Group on The Interrelationship Between Investment Flows and Technology Transfer (Geneva: UNCTAD, July 1992).

11 S. Djankov and B. Hoekman, 'Avenues of technology transfer: Foreign investment and productivity change in the Czech Republic' (Milan: Fondazione Mattei Working Paper

No. 16, 1998). Others define contractual exchanges such as licensing, joint ventures and turn key projects as arm's-length exchanges, and distinguish them from intra-firm exchanges such as FDI. See, for example, H. Pack and K. Saggi, 'Inflows of foreign technology and indigenous technological development', *Review of Development Economics*, Vol. 1 (1997), pp.81–98, quoted in Djankov and Hoekman, ibid.

12 Freeman and Hagedoorn, 'Catching up'. J. Hagedoorn, 'Strategic technology partnering during the 1980s: Trends, networks and corporate patterns in non-core technologies', *Research Policy* (1995), p.24. Pietrobelli, *Emerging Forms*.

13 Among many others, see Reddy, 'New trends', p.1,822.

14 Ibid. In 1994, the annual net income of engineers in India (Bombay) was US$2,100, in Japan (Tokyo) US$51,400, in the USA (Chicago) US$34,600, in Italy (Milan) US$23,100, in South Korea (Seoul) US$20,100, and in Brazil (Sao Paulo) US$11,600 (as reported in Reddy, 'New trends', p.1,835).

15 UNCTAD, *World Investment Report 1997* (Geneva: UNCTAD, United Nations, 1997).

16 R.E. Evenson and L.E. Westphal, 'Technological change and technology strategy', in J. Behrman and T.N. Srinivasan, *Handbook of Development Economics, Vol. III* (Amsterdam: Elsevier Science, 1995).

17 R.R. Nelson, 'Innovation and economic development: Theoretical retrospect and prospect', in J. Katz (ed.), *Technology Generation in Latin American Manufacturing Industries* (London: Macmillan, 1987), p.84.

18 Evenson and Westphal, 'Technological change', p.2,212.

19 Nelson, 'Innovation and Economic Development', p.84.

20 References on the theory of technological capabilities are: M.R. Bell, '"Learning" and the accumulation of industrial technological capacity in developing countries', in Fransman and King (eds), *Technological Capability in the Third World*, London: Macmillan, 1984), pp.187–209; H. Pack and L.E. Westphal, 'Industrial strategy and technological change', *Journal of Development Economics* (1986); J. Enos, *The Creation of Technological Capability in Developing Countries* (London: Pinter, for the ILO, 1991); J. Katz (ed.), *Technology Generation in Latin American Manufacturing Industries* (London: Macmillan, 1997); M.R. Bell and K. Pavitt, 'Accumulating technological capability in developing countries', *Proceedings of the World Bank Annual Conference on Development Economics* (Washington, DC, 1992); S. Lall, 'Technological capabilities and industrialisation', *World Development*, Vol. 20, No. 2 (1992).

21 Bell, 'Learning'.

22 A powerful way of learning is by means of training within producing firms. This has the disadvantage that training will probably stay at a level below what would be socially optimal, because of the well-known problem of incomplete appropriateness of its results. But in-firm training will be more appropriate, as the firm will provide exactly the kind and quantity of training necessary for the absorption and advancement of the specific technology; see Enos, *The Creation of Technical Capability*, p.80. Furthermore, *learning itself has to be learnt*, as it is a highly specialized process, one that involves the organization of the accumulation of technical knowledge; see J.E. Stiglitz, 'Learning to learn, localized learning and technical progress', in P. Dasgupta and P. Stoneman (eds), *Economic Policy and Technological Development* (Cambridge: Cambridge University Press, 1987).

23 R.R. Nelson, 'Research on productivity growth and productivity differences: Dead ends and new departures', *Journal of Economic Literature*, Vol. XIX (1981), pp.1,029–64. G. Dosi, 'Sources, procedures, and microeconomic effects on innovation', *Journal of Economic Literature*, Vol. XXVI, No. 3 (September 1988), pp.1,120–71.

24 Evenson and Westphal, 'Technological change'.

25   Katz, *Technology Generation*. Lall, 'Technological capabilities and Industrialisation'.
26   W.M. Cohen and D.A. Levinthal, 'Innovation and learning: The two faces of R&D', *Economic Journal*, Vol. CIX (1989), pp.569–96.
27   United Nations, *World Economic and Social Survey, 1995*.
28   Lall, 'The interrelationship between investment flows and technology transfer'.
29   In Namibia, most of the FDI used to come from South Africa until 1991, whereas in the 1990s investors have been increasingly coming from different countries.
30   A. Amsden, *Asia's Next Giant: South Korea and Late Industrialization* (New York: Oxford University Press, 1989). S. Lall, *Learning from the Asian Tigers: Studies in Technology and Industrial Policy* (London: Macmillan, 1996).
31   FDI inflows in Chile appear larger as a percentage of GDP than as a percentage of GDCF, perhaps because of the increase in the gross domestic investment ratio recorded since the early 1990s.
32   Hagedoorn, 'Strategic technology partnering during the 1980s'. Mytelka, *Strategic Partnerships and the World Economy*. Pietrobelli, *Emerging Forms of Technological Co-operation*.
33   This, like all similar databases, has a number of pitfalls, but seems to be the only reliable alternative to detailed case studies, with the purpose of giving a global picture of the phenomenon. The pitfalls are especially relevant for alliances involving LDCs' enterprises. Hagedoorn, 'Strategic technology partnering during the 1980s'.
34   Freeman and Hagedoorn, 'Catching up'.
35   Lall, *Learning from the Asian Tigers*.
36   Inter-country comparisons cannot be confidently drawn on the basis of these simple figures. Only an assessment of the growth rate of each country's MVA and labour productivity may be attempted.
37   S. Lall, 'Industrial and technological development policies in East Asia', paper presented at the International Seminar on Successful Industrial Competitiveness Policy Experiences (Santiago, Chile: organized by ECLAC, United Nations, Santiago, and the Chilean Ministry of Economic Affairs, 9–10 September 1997).

13   Nuclear Energy for Development

Gert G. Harigel

INTRODUCTION

This chapter will discuss nuclear power as an option that may satisfy increasing demands for energy. It will consider such aspects of nuclear energy as its ability to supplement dwindling fossil fuel resources, its special ability to generate electrical energy, its possible environmental desirability, its safety aspects, and its potential for nuclear weapons proliferation through the production of weapons- and reactor-grade fissile material. These issues will be evaluated for three categories of countries: developed, developing, and less-developed countries.

A detailed financial comparison of nuclear energy with other energy sources is beyond the scope of this chapter. Such a financial analysis may be of secondary importance for long-term policy decisions on future electric energy production, although that may be an idealistic view of future energy policy debates. Instead, the emphasis here will be on the need for all kinds of energy, and on their relative contributions to satisfying this need. The focus will be on electrical energy, for which nuclear energy is one of the source options. The ultimate question is whether nuclear energy technology is suitable for transfer from the developed countries which invented it to less-developed countries which would have to accept and perhaps adapt it to their needs and capabilities. Which of the less-developed countries need increasing amounts of electrical energy, which of them could operate nuclear reactors, and which have the necessary infrastructure for the effective and safe use of nuclear energy?

Any analysis of energy production requires clear distinctions between relatively abstract terms of science and the less specific (or 'popular') usage of words in policy analysis. For example, a fundamental law in physics states that energy is conserved in a closed system, that energy can never be created or destroyed, but only transformed from one form into others. This sense of 'energy conservation' should not be confused with the popular meaning of 'energy conservation', in which energy may be conserved by keeping the windows of an apartment closed, shutting off the valves on radiators, or turning off light bulbs and television sets. That popular sense of 'energy conservation' means preserving the usefulness of energy for accomplishing desired tasks by transforming it into more useful forms or preventing its degradation to less useful forms. Hence, the term 'conservation' in this context relates to preventing the transformation of energy from a more useful form to a less useful one. The creation of nuclear energy involves not the creation of energy, but the transformation of the energy stored in a nucleus into other useful forms, such as electrical energy.

It is also important for this analysis to make a distinction between renewable sources of energy. During the last several hundred million years, the energy of the sun radiated onto the Earth has converted the remains of plants and animals into fossil fuel products such as coal or hydrocarbons. Human beings are now very rapidly converting these fossil fuels, such as bituminous coal, lignite, petroleum and natural gas, into carbon dioxide, with the release of useful energy. These are called non-renewable energy sources, because over the time scale of their exploitation by technological society, these fossil fuels will not be replaced by fossilizing processes. Other sources of energy available on a more or less regular basis are solar radiation, energy contained in bio-mass, hydro-electric energy, wind and geothermal energy. These are called renewable energy sources, since they are regenerated on a short time scale. In this sense, nuclear energy is a non-renewable energy source, since it is produced from uranium, an ingredient of the Earth's crust that can never be replaced.

In practice, there are demands for various different forms of energy. Food and heat are essential energy ingredients for the survival of humans. Broadly speaking, humans everywhere have historically demanded the increasing availability of all kinds of energy, not only for survival, but also to achieve an increasing standard of living. But energy cannot be provided for this increasing demand without further technical developments. These developments require the production (transformation) of various forms of energy. The energy currently used for various essential and non-essential purposes can be conveniently subdivided into energy in the form of food, fuel, electricity, low-temperature heat for buildings and warm water, and high-temperature process heating.

For a technological society, probably the most convenient form of useable energy for many purposes is electrical energy. It represents only a fraction (about 20 per cent in industrial countries) of the total energy used, but it is very useful for heating, cooling, powering all sorts of motors, communication systems, household appliances, and so on. Yet there is much controversy about the methods for producing it. Therefore, this chapter will examine methods of production (transforming) of electric energy, analyse their advantages and disadvantages, including their compatibility with the environment, and compare nuclear energy with alternative energy production methods.

In making such an analysis, many issues must be considered, such as population growth, industrial development, availability of raw materials, arable land to produce food, forests and climate stability, and bio-diversity to balance the animal and plant environments. At the same time, safety aspects of the production of energy must be considered, including a quantitative evaluation of risks such as the possible proliferation of nuclear weapon-useable material.

This chapter will concentrate on a series of questions:

1 What will be the future trends in energy demand?
2 How much of this demand will be for electrical energy?
3 How far can renewable energy sources satisfy future energy demands?
4 Is nuclear energy better in terms of the extent of reserves, costs, safety and environmental effects than fossil fuel energy?
5 Should different standards of safety, costs and environmental concerns hold for developed, developing, and less-developed countries?
6 Is nuclear energy technology transferable to less-developed countries?

THE CHANGING WORLD

Population development

One reason for considering the expansion of nuclear reactor energy production is the rising demand for energy. The demand for energy has been increasing for several reasons, including population growth and increases in per capita energy consumption. There has been a population explosion during the twentieth century., Within the last sixty years, the world population has trebled and is now approaching six billion, as shown in Table 13.1.[1]

Forecasts are always difficult to make, and usually incorporate large errors. In 1994, the United Nations estimated a further doubling of the global population by about 150 years from now, thereafter a stabilization, and an eventual decline.[2] Others doubt that such a doubling will occur.[3] More recent estimates, based on regional growth rates, scale down the growth considerably, and due to reductions in reproduction rates, put the maximum population at 7.7 billion forty years from now.[4] A phenomenological

Table 13.1 Population data and projections

	1950	1995	2050	2150
World	2.48	5.64	9.97	11.59
Developed countries	0.73	1.02	1.17	1.36
Europe	0.56	0.73	0.77	
North America	0.17	0.29	0.40	
Latin America	0.19	0.47	0.93	0.89
Africa	0.16	0.70	2.04	2.84
Asia	1.39	3.45	5.83	7.64
China		1.21	1.90[1]	1.87
India		0.94	2.37[2]	2.90
Others		1.23	1.59	2.84

Notes:
1   Southern Asia, including China.
2   Central East Asia, including India.

Source: Klaus Heinloth, *Die Energiefrage: Bedarf und Potentiale, Nutzen, Risiken und Kosten* ('The Energy Challenge: Requirements and Potentials, Yield, Risks and Costs') (Wiesbaden: Vieweg, Handbuch Umweltwissenschaften, 1997). The data for 1950, 1995 and 2150 are from p.23, those for 2050 from p.26.

model of world population, which reproduces the past population development over the entire known history of mankind well, arrives at an asymptotic value of about 14 billion.[5] The predicted numbers are, of course, subject to large, unpredictable events, such as wars, food shortages, environmental catastrophes and epidemics.

Whereas the number of people in developed countries is predicted to remain more or less stable, or perhaps even to decline, China expects an additional 400 million inhabitants by the middle of the twenty-first century, India an additional 600 million. Other Asian, Latin American and most of the African countries also anticipate huge population increases. The last fifteen years have seen an average increase by about 80 million people per year, equivalent to adding the present population of Germany each year.[6]

PRIMARY ENERGY SOURCES

Primary energy sources are the renewable resources of coal, petroleum and natural gas. Throughout this chapter, the energy content of coal, petroleum

Table 13.2 Regional distribution of economically exploitable primary energy sources: coal, petroleum, and natural gas. Resources are in billions (giga=G) of tons of Bituminous Coal Units (tBCU)

	Total		Coal		Petroleum		Natural gas	
	GtBCU	%	GtBCU	%	GtBCU	%	GtBCU	%
North America	224.3	19.5	210.4	26.8	5.43	2.8	9.2	4.8
South & Central America	46.4	4.0	10.4	1.3	26.3	13.5	10.1	5.2
Western Europe	59.3	5.2	49.9	6.3	3.6	1.8	8.2	4.3
Eastern Europe	52.9	4.6	51.7	6.6				
Near East	182.6	15.9			128.3	65.6	61.0	33.0
Africa	85.9	7.5	61.8	7.9	11.9	6.1	13.1	6.9
Former Soviet Union	243.0	21.1	165.9	21.1	11.4	5.8	75.8	39.7
Far East	91.3	7.9	78.7	10.0	3.6	1.8	10.3	5.4
China	101.3	8.8	94.6	12.0	4.7	2.4	2.3	1.2
Australia	64.0	5.6	63.0	8.0	0.2	0.1	0.8	0.4

Source: Heinloth, *Die Energiefrage*: p.138 for the total, and pp.142, 146 and 148 for coal, petroleum and natural gas.

and natural gas will be given in tons of Bituminous Coal Units (tBCU), so that their energy contents can be compared. In terms of this unit, the total amount of proven reserves is 1,151 trillion tBCU (see the Appendix for conversion factors). Table 13.2 shows their regional distribution. These three primary energy sources are quite unevenly distributed between the continents, so that there are have and have-not countries. This uneven distribution will lead to future struggles for the possession of energy resources and for self-sufficiency.[7] These raw materials are of vital importance to international relations, and have frequently been a decisive factor in the determination of foreign policy.

Primary energy consumption

Table 13.3 shows that world primary energy consumption has increased by a factor of ten during this century. It has been increasing more rapidly than the population, reflecting an ever-increasing energy consumption per person. The most dramatic increases have occurred in the use of petroleum and natural gas.

There is a tremendous gap between the consumption per capita in various countries, with the USA at one end of the scale, and China and India at the other. The annual average per capita energy consumption in tBCU is: India, 0.3; China, 1.0; Japan, 5.7; France, 5.7; Germany, 6.0; The Netherlands, 7.3; and the USA, 11.0.[8] This represents a range of more than an order of magnitude, and the gap has been consistently widening.

Making any reliable predictions of energy consumption is therefore doubly difficult, since it must involve predictions about both population growth

Table 13.3   Worldwide energy consumption in billion tBCU, listed by various categories of energy sources

	1900	*1920*	*1940*	*1960*	*1980*	*1994*
Hydro-electric	—	—	0.10	0.28	0.62	0.84
Natural gas	—	0.1	0.2	0.53	1.84	2.61
Petroleum	—	0.2	0.4	1.59	4.32	4.53
Coal	0.8	1.4	1.6	1.80	2.60	3.08
Nuclear	—	—	—	—	0.25	0.82
Other non-commercial	0.4	0.4	0.4	0.90	1.12	1.39

Source: Heinloth, *Die Energiefrage*, p.84.

Table 13.4   Past and projected worldwide annual requirements for primary energy

	1990	*2020*	*2050*
Primary energy (billion tBCU)			
World	12.5	15.8	19.3
Industrialized countries	5.6	5.6	5.6
Developing countries	3.7	7.8	10.5
Former Eastern Bloc	3.2	2.4	3.2
Electric energy, world			
Electric energy (PWh(e))	11.5	20.0	22.0
Electric energy (thermal) (PWh(th))	34.5	60.0	66.0
Electric energy (thermal) (billion tBCU)	4.2	7.4	8.1

Note: Also shown are the demands for the electric energy sub-component, in terms of electric output, thermal energy used at 33 per cent efficiency to make that electric energy, and the tBCU equivalent of that thermal energy.

Source: Heinloth, *Die Energiefrage*, pp.88, 89.

and per capita consumption patterns. An increase of 40 per cent in global energy consumption over the next fifty years has been projected with the largest growth expected to occur in developing countries, as shown in Table 13.4.

Consequences of primary energy consumption

Running out of fuel

At the current rate of energy consumption (with a stable population and constant per capita energy consumption), the economically exploitable and proven energy reserves of coal will be exhausted about 200 years from now, of petroleum in about 50 years, and of natural gas in about 60 years (see Table 13.5).

The time-to-depletion of fossil fuel reserves will be further shortened considerably by population growth and increases in per capita energy consumption. Considering the three proven sources as a totality, and assuming that a shift in consumption from one to the other is possible, proven fossil fuel reserves as a whole could be exhausted by the year 2080.

Table 13.5   Consumption and reserves in billions of tons of Bituminous Coal Units (billion tBCU) for fossil fuels

	Cumulative consumption to 1995	Present annual consumption	Supplies	
			Proven reserves	*Estimated reserves*
Coal	180	3.1	600	7,000
Petroleum	143	4.6	200	500
Natural gas	55	2.6	165	2,300
Total fossil fuels	380	10	1,000	10,000

Source: Heinloth, *Die Energiefrage*, p.138.

In addition to the present annual consumption and proven reserves, Table 13.5 also gives data for estimated reserves. These are considerable, and could compensate for some of the growth in population and per capita consumption, but they require enormous costs to extract them from land or the ocean floor.

One alternative source of non-renewable energy involves the use of uranium (or thorium) in nuclear fission reactors to produce electricity. Uranium is relatively abundant, depending largely on the price one is willing to pay to mine it.[9] For example, the extraction of uranium from sea water, which contains 3 mg of uranium per ton, is currently being investigated. If this reservoir of about 4 billion tons of uranium could be economically exploited (a 1998 estimate for collecting the uranium on a plastic absorber was $100 per kg), it would fuel 2,000 light-water reactors for 5,000 years, and breeder reactors for 500,000 years.[10]

In the very long term, fusion reactors might provide electric energy. This would involve technologies such as plasma-magnetic confinement (*Stellerator, Tokamak,* ITER) or droplet-fusion by high-intensity laser beams or high-energy heavy ions. However, these are not likely to be available in less than half a century. Both fission and fusion energy have the advantage that they would conserve fossil fuels for alternative uses, to produce naphtha, lubricants, waxes, paraffin, asphalt, coke and plastic.

Environmental effects

The consumption of fossil fuels has a global environmental impact. Because of this increasing consumption, during the past half-century, sulphur and nitrogen emission have increased worldwide by factors of 2.4 and 4.0 respectively.[11] This has led to environmental pollution in the form of acid rain, and so on. More important is the global impact of CO_2 (carbon dioxide) emissions. During the last fifty years, world carbon emissions from burning fossil fuel have increased by a factor of 3.5, or 2.5 per cent per year. Industrial countries have increased CO_2 production by a factor of 2.3, while emissions from developing countries have increased by a factor of 25, since the latter started at a low level of fossil fuel consumption.[12]

The magnitude of the expected impact of the greenhouse gas CO_2 on the climate is still under critical discussion. However, the global average temperature increased by 0.5°C during the last hundred years,[13] most likely as a consequence of the industrial revolution. An additional increase by 1–3.5°C by the year 2100 is projected if the world stays on the present fossil fuel path, where the atmospheric CO_2 concentration is projected to reach twice the pre-industrial level as early as 2050.[14]

Renewable energies

With the potential exhaustion of primary non-renewable resources by the middle of the next century,[15] much effort is going into the development, and where possible the deployment, of solar energy, bio-mass, wind energy and geothermal power as alternative renewable energy sources.

Currently, these alternative renewable energy sources make relatively small contributions to energy supplies, except for established renewable hydro-electric power schemes, as listed in Table 13.3. Furthermore, too often the financial consequences of such renewables are not well studied, quantitative statements are not elaborated and adverse impacts on the environment are not analysed. In any case, for the present discussion, the question is: how vital is electrical energy, and how should it be produced? Renewable energy sources are not likely to be major suppliers of such electrical energy.

THE ELECTRICITY CHALLENGE

For all human activities requiring electricity, it must be produced by conversion from one of the primary or renewable energy sources. In our industrial society, it is vital to have electricity constantly available in adjustable quantities, summer or winter, day or night, for household appliances such as stoves, refrigerators, lighting, television, radio, razors, telephones, and for industries.

Electric energy production

World electrical energy production levels are given in Table 13.6 for 1971 and 1991. This shows an increase by a factor of 2.3 over these twenty years, or 4 per cent per year. Coal was the primary source for the thermal energy that is converted into electric energy; its share in the production process remained essentially constant.

Table 13.6   1971 and 1991 world electrical energy production in teraWatt-hours (TWh), arranged by source of thermal energy for production

	1971		1991	
Primary energy	*Use teraWatt-hours*	*Share %*	*Use teraWatt-hours*	*Share %*
Coal	2,142	40	4,671	39
Renewable	1,241	23	2,290	19
Nuclear	111	2	2,106	17
Natural gas	714	13	1,594	13
Oil	1,102	21	1,376	11
World	5,311	100	12,037	100

Source: Christopher Flavin and Nicholas Lenssen, *Power Surge: Guide to the Coming Energy Revolution* (New York and London: W.W. Norton, Worldwatch Environmental Alert Series, 1994), pp.61, 246.

The use of renewable energy sources, such as hydro-electric, wood, biomass, geothermal power, wind energy and photo-voltaic, almost doubled during this twenty-year period, and stayed in second place in its overall contribution to electrical energy production. During that period, there was a dramatic rise in the generation of nuclear electric power, taking it to third place, followed by natural gas and oil. However, despite a considerable rise in the use of geothermal power, wind energy and photo-voltaic, these three renewable energy sources contribute only on about 1 per cent to global electricity production. Although the number of new nuclear reactors is decreasing, the world electricity-generating capacity of nuclear power plants has still been rising, due to improved performance. The current rate of growth of that capacity is about 1 per cent per year.[16]

Heinloth reports that in 1995 the world's electrical power consumption of 12.5 PWh came 38%, 11%, 14%, 18% and 19% from coal, petroleum, natural gas, nuclear and hydro-electric power stations respectively. He presents one model of how a future annual demand of 20–25 PWh in 2050 might be satisfied: 30–44%, 15%, 15%, 10%, 5% and 25% from fossil fuel, nuclear, bio-mass, solar, wind and hydro-electric sources respectively.[17]

Distribution of electrical energy consumption

The consumption of electrical energy is unevenly distributed among countries, both in terms of per capita consumption of electricity, and in the fraction of energy for a given country that is electrical. The countries of the world are often classified, according to their economy and their living standard, into three major groups: developed countries, developing countries, and less-developed countries. To illustrate that distinction, Table 13.7 lists countries by these three categories, with an indication of their status regarding nuclear power. Table 13.8 shows that these groups differ greatly in their energy consumption patterns. The less-developed countries use much less energy per capita, consume a smaller fraction of that energy in the form of electrical energy, and produce less of that electrical energy through nuclear power reactors.

Energy consumption patterns

To understand more clearly the different energy consumption patterns for the various stage of development among nations, the energy situation will be evaluated for a few selected countries and regions from each of these groups. This will establish a framework for analysis, allow some comparisons, and permit some preliminary conclusions to be drawn.

Energy consumption patterns for Germany

The energy requirements of the industrialized country of Germany will help in an analysis of whether nuclear energy is absolutely necessary, is just a convenience, or could be completely abandoned. Germany has a high living standard and a stable population, and comprehensive data is available for it, allowing a thorough analysis. Energy requirements will be subdivided according to the scheme used by Klaus Heinloth.[18]

Food   A requirement of 3,500 kilocalories per person per day (for an explanation of units, see the Appendix) is necessary to keep the population

Table 13.7 A list of developed, developing, and less-developed countries and their nuclear power status

---

Developed Countries
Albania, **Armenia**, Austria, Australia, Azerbaijan, Bahamas, Belarus, **Belgium**, Bermuda, **Bulgaria**, Bhutan, Bosnia-Herzegovina, **Canada**, Croatia, **Czech Republic**, Denmark, Estonia, **Finland**, **France**, Georgia, **Germany**, Greece, Greenland, Grenada, **Hungary**, Iceland, Italy, Ireland, Israel, **Japan**, **Kazakstan**, Kyrgyzstan, Latvia, **Lithuania**, Luxembourg, Macedonia, Maldives, Moldova, **Netherlands**, New Zealand, Norway, Poland, Portugal, **Romania**, **Russia**, **Slovak Republic**, **Slovenia**, **Spain**, **Sweden**, **Switzerland**, Tajikistan, Turkey, Turkmenistan, **Ukraine**, **UK**, **USA**.

Developing Countries
Algeria, Angola, **Argentina**, Bahrain, Barbados, Bahamas, Belize, Bolivia, **Brazil**, Brunei, Cameroon, Cambodia, Chile, **China**, Columbia, Congo, Costa Rica, Côte d'Ivoire, *Cuba*, Cyprus, Dominica, Dominican Republic, Ecuador, *Egypt*, Fiji, Gabon, Ghana, Guatemala, Guyana, Honduras, **India**, Indonesia, *Iran*, Iraq, Israel, Jamaica, Jordan, Kenya, *North Korea*, **South Korea**, Kuwait, Lebanon, Libya, Madagascar, Malaysia, Mali, Marshall Islands, Mauritius, **Mexico**, Federated States of Micronesia, Mongolia, Morocco, Namibia, Nicaragua, Oman, **Pakistan**, Panama, Papua New Guinea, Paraguay, Peru, Philippines, Qatar, St Vincent & the Grenadines, El Salvador, Saudi Arabia, Senegal, Seychelles, Singapore, Solomon Islands, **South Africa**, Sri Lanka, Surinam, Swaziland, Syria, **Taiwan**, Thailand, Tonga, Trinidad & Tobago, Tunisia, Tuvalu, United Arab Emirates, Uruguay, Venezuela, Vietnam, Zaire, Zambia, Zimbabwe.

Less-developed countries
Afghanistan, Bangladesh, Benin, Botswana, Burkina Faso, Burundi, Cape Verde, Central African Republic, Chad, Comoros, Djibouti, Equatorial Guinea, Eritrea, Ethiopia, Gambia, Guinea, Guinea-Bissau, Haiti, Kiribati, Laos, Lesotho, Liberia, Malawi, Mali, Mauritania, Mozambique, Mianmar, Nepal, Niger, Rwanda, Samoa, Sierra Leone, Somalia, Sudan, Tanzania, Togo, Uganda, Vanuatu, Yemen, North Yemen, South Yemen.

Note: Countries with nuclear power stations are marked as follows: **bold** indicates countries with operational nuclear power stations; ***bold plus italics*** indicates countries with nuclear power stations under construction; *italics* indicate countries with nuclear power stations in the planning stage. No less-developed countries fall into any of these nuclear categories (American Physical Society (APS), available at: <http://aps.org/public_affairs/popa/popaii-2.html>).

Source: SIPRI Yearbook, *World Armament and Disarmament 1995* (Oxford: Oxford University Press, 1995), pp.510, 511.

Table 13.8 **Population energy consumption per capita of coal, natural gas, oil and hydroelectric (wood not included), electric energy consumption per capita, and usage of nuclear electric energy, for countries grouped according to their stage of development, as listed in Table 13.7**

	Countries		
	Developed	*Developing*	*Less-developed*
No. of countries	46	87	41
Population (million)	1,375.2	3,928.1	515.0
% of world population	23.6	67.5	8.8
Per capita energy consumption (tBCU/y)	5.633	1.009	0.170
Per capita coal (tBCU/y)	1.880	0.480	0.002
Per capita gas (tBCU/y)	1.591	0.130	0.004
Per capita oil (tBCU/y)	1.877	0.344	0.031
Per capita hydro (tBCU/y)	0.285	0.055	0.133
Per capita electricity consumption (kWh/y)	6.360	830	54
Per capita electricity consumption (tBCU/y)	0.782	0.102	0.007
No. of countries with nuclear power	23	9	0
No. of nuclear reactors	395	38	0
Nuclear electric power (GW(e))	323	22.5	0
Nuclear/total electric energy (%)	23.8	4.4	0
% of world nuclear electricity	94	6	0

Note: Energy data are mainly for 1994.

Source: *1998 Britannica Book of the Year* (London: Encyclopaedia Britannica, 1998), pp.756–61, 820–5.

healthy; this amounts to 1.4 EJ/year. The agricultural bio-mass from 47 per cent of the country's area produces 1.6 EJ/year, resulting in a small overproduction. In 1995, 12 per cent of the grain area in the European Union was kept out of production, but this was reduced to 5 per cent in 1997,[19] including the area used in Germany, because of a decrease in worldwide carry-over food stocks, approaching a food stockpile sufficient for only fifty days.[20] More land will be needed in the future for agricultural use, making less available for other purposes.

Fuel The requirement is 60 million tons of gasoline per year, which corresponds to 2.6 EJ/year, with 70 million tons of petroleum available per year.

Germany will never be independent in terms of oil imports, since it can never produce enough bio-mass for conversion into fuel. The amount of bio-mass equivalent to this oil demand is 5 EJ/year; this would have to be harvested from 1.5 times the area of Germany. The automobile fleet is approaching an asymptotic value in Germany, and most industrial countries; however, the worldwide automobile fleet is expected to double within the next twenty-five years, and will consume one-third of all oil production.

Electricity   The requirement is 530 TWh per year. This quantity could come from plants burning 170 millions tons of bituminous coal. It could also come from the fission of 12,000 tons of natural uranium in nuclear power plants. It could not come solely from hydro-electric power stations, since the equivalent of 40 fully exploited Rhine rivers would then be needed. If solar photo-voltaic cells were to be used, 7,000 km^2 of solar cells would be required, combined with the storage capacity of 8 billion tons of batteries, or stored in reservoirs (each with the physical characteristics of the *Edertalsperre*) covering a total area the size of the former West Germany. Only the first two energy options, coal and nuclear, can provide a major fraction of Germany's electrical energy needs, although hydro- and solar-electric production methods should be exploited as much as possible to feed some supplementary energy into the electric grid.

Heating   Buildings and hot water (less than 100°C) consume 3 EJ year. This requirement is presently covered by a combination of burning bituminous coal, petroleum and natural gas. To provide this energy by burning wood would require an annual consumption of 450 million of m^3 wood. This would have to be harvested from a surface 2.5 times the area of Germany. A hypothetical case can be made for combining 2,000 km^2 of flat solar collectors with 20 km^3 of hot water storage.

High-temperature process heat   This requires 1.8 EJ year, which is larger than the demand for food energy. It could hypothetically be provided by burning 80 millions tons of charcoal or 300 million m^3 of wood per year, which would require 1.5 times the area of Germany. If this demand were to be covered by the use of electrical energy, 1.9 EJ would be required, which would double the demand for electrical energy.

The data indicate that importing fossil fuel products remains a necessity for Germany, since no (complete) replacement for it from other sources can be envisaged under any reasonable assumptions. Renewable energy cannot be expected to satisfy more than a few per cent of the energy demand. Alternative energies require a lot of land surface. Conservation in the form of better-insulated buildings, reducing the temperatures in living and working spaces, disconnecting unnecessary electric equipment and designing more efficient conversion engines should become a high priority to at least slow the increase in energy demand. However, nuclear energy is still needed at the present level to supply about 33 per cent of Germany's energy needs for many years to come, and the number of power stations might even be increased.

Energy for large, rapidly developing countries

China and India will serve to illustrate the energy consumption and production patterns of rapidly developing countries. In considering the data, it is necessary to keep in mind the relative populations of China, India, Germany and the USA: 1,219 million, 953 million, 82 million and 265 million, respectively.

Both of these large developing countries face a population explosion, rapid industrialization, and substantial annual losses of arable land. China experienced a cropland loss of 3 per cent from 1986 to 1992[21] through the growth of cities, the building of roads, and the expansion of deserts. It also experienced dwindling aquifers, and a decrease in forest area of less than 1 per cent per year.

China's energy problems are aggravated by its population explosion and the concentration of most of its major cities and industries in coastal areas. China's coal reserves are in the north-east and central northern parts, far away from the places where energy is needed. The underdeveloped transport system is inadequate to move sufficient quantities of coal. Hydro-electric power is in the south-west, again relatively far from the places where it is most needed.[22] It is well known that transmission of electricity over large distances results in considerable losses.

The dwindling acquifers aggravate the situation, both for drinking water supplies and for irrigation water. In general, the operational desalination plants are where the high cost of desalinated water can be afforded. A population can usually afford to pay about ten times as much for water for domestic purposes as for agricultural water. Proposals for large-scale nuclear desalination facilities promise to lower the cost of desalinated water to a level that most industries and a few agricultural enterprises can afford. This can be considered an additional argument in favour of having nuclear power plants close to China's coastline.

Similar conditions and requirements exist for the Indian subcontinent. The concentration of rapidly growing major cities requires large amounts of electrical energy, which can be provided by renewable energies only on a tiny scale. Again, nuclear power might become the preferred option.

China and India have put considerable effort into the improvement of scientific education, which has resulted in a remarkable number of

Table 13.9   Energy production and consumption patterns for Germany, China, India and the USA in 1994

	Produced	*Consumed*
Germany		
Electricity	528,221,000,000 kWh	530,558,000,000 kWh
Hard coal	57,632,000 tons	66,255,000 tons
Lignite	207,077,000 tons	209,308,000 tons
Crude petroleum	21,535,000 barrels	793,500,000 barrels
Petroleum products	99,578,000 tons	113,839,000 tons
Natural gas	20,904,000,000 m^3	92,770,000,000 m^3
China		
Electricity	928,083,000,000 kWh	926,037,000,000 kWh
Coal	1,239,902,000 tons	1,231,928,000 tons
Crude petroleum	1,069,320,000 barrels	1,024,375,000 barrels
Petroleum products	106,629,000 tons	114,972,000 tons
Natural gas	17,540,000,000 m^3	17,540,000,000 m^3
India		
Electricity	351,000,000,000 kWh	385,902,000,000 kWh
Coal	273,859,000 tons	284,497,000 tons
Crude petroleum	244,743,000 barrels	434,149,000 barrels
Petroleum products	43,575,000 tons	56,722,000 tons
Natural gas	17,638,000,000 m^3	17,638,000,000 m^3
US		
Electricity	3,268,250,000,000 kWh	3,312,888,000,000 kWh
Coal	937,580,000 tons	843,873,000 tons
Crude petroleum	2,464,000,000 barrels	5,024,000,000 barrels
Petroleum products	704,201,000 tons	737,681,000 tons
Natural gas	530,014,000,000 m^3	592,209,000,000 m^3

high-quality scientists. In both countries, there are probably enough qualified scientists and technicians to operate nuclear power plants safely.
Educational standards are superior to those in most of the other developing
countries.

As the standard of living increases in China, food consumption increases, possibly leading to a severe food shortage in the future. Its grain

lands decreased over the last forty years by 50 per cent to 0.07 hectares per person, and may fall by a further 50 per cent by the year 2030. While grain production is still stable in India at 180 kg per person/year, it already appears to be dropping slightly in China from the present 290 kg per person/year.[23]

Both countries face the problem of where to find increasing energy resources, particularly electrical energy. These energy resources are required to satisfy the demand for better transportation and growing industries. They could also be useful in the production of drinking water and water for irrigation by desalination, which might become necessary due to the dramatic drop of aquifer levels. The growth of the automobile fleet also needs to be accommodated. While this fleet is presently still modest in size, increasing from about 1 million in 1990 to the present 2.7 million, the transportation ministries expect it to grow by almost an order of magnitude by the year 2010, to reach about 22 million cars.[24]

China has 10 per cent of the world's fossil resources, including 12 per cent of the coal, 2.4 per cent of the oil, and 1 per cent of the natural gas. China's coal consumption is four times that of Germany, has risen by 50 per cent during the last decade, and rose by 4.5 per cent in 1996.[25] Even though China's electricity production is presently only 30 per cent more than that of Germany, the use of coal on a large scale in power stations without adequate removal of hazardous particulate matter has created a severe pollution problem. Nuclear power plants produce only 1 per cent of electrical energy, the three nuclear reactors in operation generating 2,100 MW. However, nuclear electrical energy is intended to increase to 20,000 MW by the year 2010, using two nuclear reactors that are under construction and a further four that are at the planning stage. China supposedly plans to have about 240 nuclear power reactors by the year 2050.[26] Even this number of reactors would satisfy only 13 per cent of China's electricity needs by then.

In India, electricity production is two-thirds that of Germany, with 5 per cent of it nuclear, while total coal consumption equals that of Germany. Since India possesses only limited fossil fuel resources and faces a huge population increase, it may have to opt for greater production of nuclear electricity.

In general, China and India, representing one-third of the human population, have until now made little use of the nuclear sector, which averages 17 per cent world wide. Table 13.10 gives the present electricity production according to production means in China and India, as well in two other South and East Asian countries bordering each other. An increase in the use of hydro-electric power stations may be possible in principle, but such major projects may lead to international political struggles over flood control, as well as the availability of drinking and irrigation water downstream

Table 13.10   Relative share of electricity production of different production means in selected South and East Asian countries

	Hydro-electric %	Main rivers	Fossil %	Nuclear %
Pakistan	34.0	Indus	65.1	0.9
India	18.5	Ganges	80.0	1.5
China	18.1	Yangtze	80.4	1.5
Bangladesh	6.0	Ganges, Brahmaputra	94.0	0

Source: *1998 Britannica Book of the Year* (London, Encyclopaedia Britannica, 1998), pp.820–5.

along the big rivers. The Three Gorges Dam on the Yangtze river is planned to produce 18,200 MW of electrical energy, which corresponds to 18 one-Gigawatt nuclear power stations. It is scheduled to be operational by the year 2009. It will require a 600 km-long reservoir, and will take up 970 km^2 of cropland. Of the 9,573 million km^2 of China's land surface, only 10 per cent is considered to be cultivatable, so the dam will reduce fertile cropland by 0.1 per cent. Arguments in favour of building the dam, apart from 'clean' electricity production, include the control of floods and protection of humans. Among the detrimental effects, there is a substantial danger of damage by earthquakes, with catastrophic consequences.

Energy for less-developed countries

The category of less-developed countries includes all Sub-Saharan countries close to the equator. This implies that there is less need for energy for heating purposes, but electrical energy for cooling is highly desirable. Basically, for these countries the competition is between the use of scarce renewable wood resources to meet minimal food, heat and transport requirements, and the need to use these to improve industry and the supporting infrastructure. For example, increased use of scarce firewood for cooking is threatening to overexploit the land, turning it into desert.

Could these countries really use substantially more electricity in an effective way? If so, huge investments would be necessary to construct an electricity grid and maintain it. If electrical energy could indeed lead to major improvements for the less-developed countries, is nuclear power the right solution? There is doubt that countries with a low level of technical and scientific education can operate and maintain nuclear reactors safely. A serious accident is more likely in such countries. When we ask whether technology can make a contribution toward solving energy problems in the less-developed countries, the question is not how much petroleum can be imported, but might solar electric energy, produced in small local installations or solar-operated stoves, be a better solution? If these countries do need more electrical energy, would it not be better to transport it to them from developing or developed countries?

THE CHALLENGE OF NUCLEAR ELECTRICITY

If a developing or less-developed country decides that it requires more electrical power in the future, then it will evaluate nuclear energy as one potential source. In evaluating nuclear power, economic aspects will obviously have to be taken into account. But nuclear power has been challenged on three grounds: the safety of nuclear reactors, the risk of nuclear proliferation, and the disposal of nuclear waste.

Nuclear experts generally claim that improved nuclear technologies promise a very good safety record. Nevertheless, it may be useful to review briefly the past performance of nuclear reactors, and this performance record should be compared to accident and health effects caused by using or producing electricity from other energy sources, such as those that occur during coal mining, the collapse of hydroelectric dams and so on. A detailed, quantitative comparison of all possible risks of all energy sources is difficult, but the risks that the public is willing to accept in using other forms of energy, such as in transportation, place nuclear energy into perspective.

Nuclear accidents

At the end of 1996, there were 433 nuclear reactors operating, producing 345.5 Gigawatts of electric power (GW(e)). Another 45 reactors were under construction to give a further net output of 38.7 GW(e). The net annual generation of the existing reactors is 2 228 TW-hours, or 254 GW-years, corresponding to 17 per cent of the worldwide electricity production. Table 13.11 gives data for each country possessing nuclear power reactors.

Nuclear power reactors can be grouped into light-water reactors (LWRs), heavy-water reactors (HWRs), gas-cooled graphite-moderated reactors (GGRs), graphite-moderated light-water reactors (RBMKs), high-temperature reactors (HTRs) and breeder reactors (BRs). The dominant

Table 13.11   The number of nuclear power reactors that were operating and under construction at the end of 1996, their total generating capacity, the net nuclear power generated and the nuclear share of the electricity generated in 1995

Countries	Operating (end 1996)		Under construction (end 1996)		Net generation (1995)		% nuclear (1995)
	No.	*GW(e)*	*No.*	*GW(e)*	*TWh*	*GWy*	
Developed							
USA	109	100.5			673	76.9	22
France	55	57.4	4	5.8	359	40.9	76
Japan	51	41.0	5	5.7	287	32.8	33
Germany	20	22.2			154	17.6	29
Russia	26	19.8			99	11.3	12
Canada	22	15.4			92	10.5	17
UK	35	12.7			78	.8.9	25
Sweden	12	10.1			67	7.6	47
Ukraine	14	12.1	4	3.8	66	7.5	38
Spain	9	7.2			53	6.1	34
Belgium	7	5.5			39	4.5	56
Switzerland	5	3.1			24	2.7	40
Finland	4	2.3			18	2.1	30
Bulgaria	6	3.4			17	2.0	46
Hungary	4	1.7			13	1.5	42
Czech Republic	4	1.6	2	1.8	12	1.4	20
Slovak Republic	4	1.6	2	0.8	11	1.3	44
Lithuania	2	2.8			10.6	1.2	86
Slovenia	1	0.6			5	0.5	40
Netherlands	2	0.5			4	0.4	5
Kazakhstan	1	0.1			0.1	0	0
Romania	1	0.7	1	0.6	0	0	0
Armenia	1	0.4			0	0	0
Developing							
Korea, South	11	9.1	9	7.6	64	7.3	36
Taiwan	6	4.9	2	2.7	34	3.9	29
China	3	2.1	6	4.6	12	1.4	1
South Africa	2	1.8			11	1.3	6
Mexico	2	1.3			8	1.0	6
Argentina	2	0.9	1	0.7	7	0.8	12
India	10	1.7	6	1.7	6	0.7	5
Brazil	1	0.6	2	2.5	2.5	0.3	1
Pakistan	1	0.1	1	0.3	0.5	0.1	1
World Total	433	345.5	45	38.7	2,228	254	17

Source: APS: <http://aps.org/public_affairs/popa/popaii-2.html>.

reactor groups are LWRs, HWRs and GGRs. Breeder reactors are intended to produce more fissile plutonium than they consume in fissile uranium-235. A detailed discussion of the various reactor type is beyond the scope of this chapter. However, HWRs of the CANDU (Canadian Deuterium) type may be particularly interesting for developing countries, since they can be operated with natural uranium, which is more easily accessible and afford-able for them. But HWR pressure vessels have a volume about a factor of ten larger than those of other reactor types, and therefore require more initial investment.

Until now, there has been only one major civilian nuclear power reactor catastrophe with considerable loss of human life: Chernobyl in the Ukraine. The radioactive releases from accidents at Sellafield in the UK, Chelyabinsk-40 in Russia and Three Mile Island in the USA all had relatively little impact on the health of populations. The Chernobyl reactor melt-down of the core and the radioactive releases in the subsequent chemical explosion will – by some estimates which are in reasonable agreement with each other – result over a period of fifty years in about 20,000–30,000 persons being affected by cancer.[27] Better estimates of the health effects of this disaster will be only be known after decades, when the results of a systematic follow-up of those people who were probably exposed to radiation doses significantly higher than background are evaluated from the studies of the Nuclear Safety Institute of the Russian Academy of Sciences in Moscow.[28]

This important release of radioactivity into the environment happened at an outdated RBMK reactor due to disregard of operating procedures, and was aggravated by the absence of a secondary containment building. Only 16 RBMKs for electricity production exist in the former Soviet Union, and these reactors were also used to make plutonium for nuclear weapons pro-grammes.

However, the risks from LWRS or HWRS are much lower than for RBMKS. The probability of a serious accident for LWRS of the modern type has been estimated as about 10^{-6} per reactor-operating year. For the 20 LWRs in Germany, this would be one serious accident during 50,000 years of operation: an accident rupturing the containment building and killing more than 1,000 people. For the 330 LWRs in operation worldwide, such a 1,000-fatality accident has been estimated to occur once every 3,000 years.[29]

These risks of nuclear accidents are small compared to the risks of other types of fatal accidents. For instance, in Germany during 1993, there were about 400,000 car accidents resulting in personal injuries, about 130,000 persons were severely injured, and about 10,000 deaths were recorded.[30] For comparison, the main risks of fatal accidents per person and per year in the USA are shown in Table 13.12. The estimated risk from a nuclear reactor accident is much smaller than those from all other accidents.

Table 13.12   Mean risk of fatal accident for different causes in the USA, 1986

Type of accident	No. per year	Probability per person per year
Car accident	55,791	1/4,000
Fall	11,827	1/20,000
Fire and hot substances	7,451	1/25,000
Flooding	6,181	1/30,000
Firearms	2,309	1/100,000
Plane crashes	1,778	1/100,000
Struck by a falling object	1,271	1/160,000
Electrocution	1,148	1/160,000
Lightning	160	1/2,000,000
Tornado	91	1/2,500,000
Thunderstorm	93	1/2,500,000
All accidents	111,992	1/1,600
Nuclear power accidents (for 100 reactors)		1/5,000,000,000 (very uncertain)

These more conventional risks are largely accepted by the population as the cost of improved living standards, as part of modern life. For example, the losses of life in coal mines, and the health problems due to inhalation of coal dust, are accepted by the public as one cost of having electricity. Yet the number of cases of pneumoconiosis from the 1930s to the mid-1950s in Germany alone amounted to about 30,000,[31] and 546 people lost their lives worldwide in mining accidents in 1997.[32] In contrast, the words 'atomic' or 'nuclear' create a perception of a larger risk – an estimate that cannot be justified on the basis of a straightforward risk analysis.[33]

The debate about the acceptability of nuclear reactors generally revolves around the appropriateness of a purely quantitative risk analysis. Many objections are raised against the use of risk comparisons, for example those of car accidents versus those of nuclear power. One argument is that the use of a car is a free choice, while nuclear power is imposed on the population by their government. However, if a pure risk assessment or cost-benefit calculation is made in purely quantitative terms, nuclear power is usually evaluated as much less risky, and with a much better cost-benefit ratio. It is argued that it may be better to replace a risk analysis with a comparison of cost-benefit ratios: lives lost per MW-year of electricity produced, versus the lives lost per passenger-mile, for example – where each risk and benefit would have to be ultimately measured by the same parameters, such as the dollars in time saved and convenience, profit per mile driven, and the dollar value of a human life saved or lost.

Nuclear proliferation

A second unique feature of nuclear power is its potential to increase the risk of nuclear proliferation. Indeed, nuclear power reactors produce fissile plutonium: each 1,000 MW(e) PWR produces about 200 kg of reactor-grade plutonium annually. The majority of scientists now accept the idea that reactor-grade plutonium can be used for building nuclear weapons, albeit with a reduced explosive yield and with some other inconveniences compared to weapons-grade material. Thus, the potential for nuclear proliferation may be increased if more nuclear power reactors are being built. However, the potential for nuclear proliferation also exists because of the large quantities of weapons-grade material that has been produced and stored. The stockpile of weapons-grade plutonium (Pu-239) and highly enriched uranium (HEU, enriched to more than 20 per cent in U-235) are 228 tons and 1,722 tons respectively,[34] in each case about a third is inside weapons, stored at secret locations, or declared excess. On average, only 4 kg of Pu-239 or 15 kg of HEU are needed to build one weapon with the explosive power of the Hiroshima bomb. The quantities of weapons-grade material are so large, that the risks of proliferation by diversion or theft appear to far exceed the risks of diversion from additional nuclear power reactors in developing or less-developed nations.

Nuclear waste

The problem of the safe storage of nuclear waste is a third seemingly unique feature of nuclear power. However, experts argue that the problem of nuclear waste is soluble, and that in any case the quantities of nuclear waste are small (in terms of both volume and toxicity) compared to the quantities of other toxic wastes generated by other technologies, including other energy technologies. Table 13.13 shows the annual production of waste in kilogram per capita in France, a country with one of the largest numbers of nuclear reactors. In terms of waste classified as highly toxic, nuclear waste represents only 1 per cent. In any case, disposal of nuclear waste is in most cases handled by developed countries, which carry out the reprocessing of

Table 13.13   Annual production of waste per inhabitant in France

Type of waste	Quantity per person (kg/year)
Household waste (kitchen garbage, diverse domestic scrap, etc.)	360
Agricultural waste (plastic, farming scrap, etc.)	7,300
Industrial waste (metal waste, iron, non-iron, powders, technological waste)	3,000
part thereof classified as toxic	100
Hospital waste (plastics, cellulose, etc.)	15
Nuclear waste (packaged)	1.2
From CEA (defence part)	0.25 (0.08)

Source: Charpak and Garwin, *Feux Follets*, p.189.

the commercial nuclear reactor cores. Thus, nuclear waste is not a major problem for nuclear reactor installations in developing and less-developed countries.

CONCLUSIONS

Energy supplies are intrinsically related to the well-being and survival of humans all over the globe, and with their ability to live together in peace. Therefore, a major challenge to all countries is to provide sufficient energy resources to their people. The question here is: how much nuclear electric energy is needed in the mix of energy for countries at different levels of development? Are nuclear power reactors an appropriate solution, not only for developing countries but for less-developed countries? Is a transfer of nuclear reactor technology to less-developed countries appropriate?

The most important social challenge is to guarantee the production of enough food and the provision of safe drinking water in order to minimize the number of deaths caused by starvation and disease. At present, 18 million people per year are victims of these food and clean-water shortages, caused not primarily by production limitations, but rather by inadequate distribution of food due to financial considerations, the unavailability of safe water, and inadequate medical care. This provision of food and water is not directly linked to energy supplies. In fact, the desire for more energy can interfere with satisfying these basic needs. For example, it seems inadvisable to use agricultural land for the production of bio-mass fuels for transportation and electrical energy.

The second task – no less challenging – is to narrow the gap in the standard of living between rich and poor countries. This implies that sufficient energy has to be available in all countries to encourage structural improvements and technological developments in industry and increases in living standards. Electrical energy can assist in technological development, as well as having the advantages of convenience in all fields of life, be it in powering engines for mass transportation, machinery in factories, illumination, heating, household equipment or information systems (radio, television, telephone). The latter, for example, offers direct improvements in both social structures and living standards through help with better education and improved social interaction. Thus, electrical energy is vital for closing the gap between developed and less-developed countries.

However, electrical energy cannot be considered in isolation from broader energy demands and supplies. It has to be seen in the context of the variety of total energy requirements. Each nation and region has to make its own decisions as to what fraction of its total energy should be contributed by electricity. Within this context of need for electrical energy, the next decision must be how much of that electrical energy should be provided by nuclear energy. Despite a reasonably good performance by nuclear power stations, there is a strong anti-nuclear movement in several, mainly industrialized countries. However, none of the production mechanisms of electrical energy are free from adverse side effects, and none are absolutely safe. The preceding data suggest that the world will have to live with nuclear energy for the near and probably more distant future, or else may have to drastically reduce the consumption of electrical energy.

Which countries would find nuclear power most suitable? Future building of nuclear reactors in developing countries, such as China and India, can be justified on ecological grounds, and may well be recommended, provided some preconditions can be fulfilled:

1 Only modern, safe and cost-effective nuclear reactors should be installed. For instance, HWRs of the CANDU type have a high initial construction cost, but provide the important advantage of using natural uranium. This avoids the costly enrichment process and dependence on delivery of enriched uranium-235 from developed countries. The absence of enrichment plants in the developing countries would reduce the danger of clandestine accumulation of weapons-grade material.

2 Accident and proliferation risks must be minimized by using operators who are well trained and supervised, perhaps even employed by the International Atomic Energy Agency in Vienna. A second reactor

accident like that at Chernobyl, whether caused by human operator mistake or equipment failure, would deal a death blow to the production of nuclear power all over the world.

3 The radioactive waste problem must be addressed at an early stage.

The situation for less-developed countries is far more complex. These countries also need an increase in electricity supplies, but they have the additional problem that the effectiveness of a nuclear power system must first be established. It will be much more difficult to afford modern, safe and cost-effective reactors, accident risks may be higher because of a weak infrastructure, and the radioactive waste problem cannot be solved internally. Given the weak infrastructure of most of these less-developed countries, it might be better to start with imported electricity, wherever possible.

In general, decisions about nuclear power must be made in the context of an overall energy policy. Long-term political and economic decisions have to be made, the sooner the better. Energy priorities must be defined to avoid wars over the possession of energy resources, such as access to oil wells. The production of any form of energy should not be achieved by sacrificing food production from arable land for secondary amenities, neither in the developed, developing nor the less-developed world.

The pros and cons of renewable energies should be evaluated carefully, and their impact on the environment and bio-diversity studied. Only then can the need for nuclear power be evaluated properly. There is concern among experts that insufficient knowledge of basic physics, statistics and comparative risk factors has led to an unjustifiably strong anti-nuclear movement. These studies should be performed for each individual country or region with similar depth to the analysis by Heinloth,[35] and as outlined in this chapter in the case of Germany. A comprehensive analysis of the economic, social, and environmental energy situation must be carried out before it is possible to judge the desirability of transferring nuclear technology to less-developed countries.

APPENDIX

Prefixes

mega (M) = 10^6, giga (G) = 10^9, tera (T) = 10^{12}, penta (P) = 10^{15}, exa (E) = 10^{18}.

Conversions

Note: Some of the data cited in this chapter originate in the USA, where the Imperial ton (=2,000 lb) is still used, while other data from the rest of the world are expressed in tonnes (1,000 kg). Standardizing these units for such large figures would have been unwieldy, so the ton/tonne distinction is correct in each instance.

1 ton is equivalent to 0.907 tonnes
1 exajoule (EJ) = 10^{18} J
1 terawatt (TW) = 10^{12} watts = 31.5 exajoules per year = 31.5 EJ/year
1 terawatt-year-thermal (TWy(th)) = 31.5 EJ = 1 billion tons of bituminous coal
1 terawatt-hour (TWh) = 1 billion kilowatt-hours (billion kWh)
1 ton BCU (tBCU) = 8.14×10^3 kilowatt-hours (thermal) = 2.7×10^3 kWh (electric)
1 ton BCU = 695 m^3 of natural gas, 0.70 tons of oil, or 5.11 barrels of oil
1 acre = 4,046 m^2

REFERENCES

1   The author is grateful to Professor Klaus Heinloth for permission to use many of the tabular and graphical presentations from his book *Die Energiefrage: Bedarf und Potentiale, Nutzen, Risiken und Kosten* ('The Energy Challenge: Requirements and Potentials, Yield Risks and Costs') (Wiesbaden: Vieweg, Handbuch Umweltwissenschaften, 1997).
2   United Nations, *World Population Prospects* (New York: United Nations, 1993, 1994).
3   Wolfgang Lutz, Warren Sanderson and Sergei Scherbow, 'Doubling of world population unlikely', *Nature*, Vol. 387 (19 June 1997), pp.803–5.
4   Charles Krauthammer, 'Immigration: America's cure for the "birth dearth"', *International Herald Tribune*, 18/19 July 1998 quoting a study by Nick Eberstadt in *Public Interest* (Fall 1997).
5   Sergei Kapitza, 'The phenomenological theory of world population growth', *Physics-Uspekhi*, Vol. 29 (1996), pp.57–61.
6   Lester L. Brown, Michel Renner and Christopher Flavin, *Vital Signs 1997: The Environmental Trends that are Shaping our Future* (New York: W.W. Norton, for the Worldwatch Institute, 1997).
7   Arthur H. Westing (ed.), *Global Resources and International Conflict: Environmental Factors in Strategic Policy and Action* (Oxford: Oxford University Press, for SIPRI, 1986).
8   Heinloth, *Die Energiefrage*, p.82.
9   Ibid., p.269.
10 Tadao Seguchi, Director of Material Development at the Japan Atomic Energy Research Institute, quoted by Richard L. Garwin in 'The Nuclear Fuel Cycle' (Paris: draft for discussion, 4 December 1998): 'LWR fuel elements from natural uranium at $700 per kg are cheaper than MOX from reprocessing of spent fuel.'

11 Christopher Flavin and Nicholas Lenssen, *Power Surge: Guide to the Coming Energy Revolution* (New York and London: W.W. Norton, Worldwatch Environmental Alert Series, 1994).

12 Ibid., p.246.

13 Brown et al., *Vital Signs 1997*, p.61.

14 Lester L. Brown, Christopher Flavin and Hilary French, *State of the World 1999* (New York, London: W.W. Norton, 1999), pp.14, 15.

15 Brown et al., *Vital Signs 1997*; Brown et al., *State of the World 1999*.

16 Brown et al., *Vital Signs 1997*, p.49.

17 Heinloth, *Die Energiefrage*, p.400.

18 Ibid., pp.486–9.

19 Brown et al., *Vital Signs 1997*, p.40.

20 Ibid., p.35.

21 Lester L. Brown, *Who Will Feed China? Wake-up Call for a Small Planet* (New York, London: W.W. Norton, Worldwatch Environmental Alert Series, 1995).

22 *Encyclopedia Britannica*, Vol. 16 (London: Encyclopaedia Britannica, 1988, 15th edn), p.54.

23 Brown, *Who Will Feed China?*, p.29.

24 Ibid., p.58. *1997 Britannica Book of the Year* (London: Encyclopaedia Britannica, 1997); pp.557, 612, 627, 741.

25 Brown et al., *Vital Signs 1997*, p.46.

26 Garwin, 'The Nuclear Fuel Cycle', p.6, quoting information from the Committee on International Security and Arms Control, National Academy of Sciences, US–Beijing discussions of 1997.

27 Heinloth, *Die Energiefrage*, p.260. Richard L. Garwin, 'More on deaths due to Chernobyl', *Physics and Society*, Vol. 28, No. 1 (January 1991), p.2.

28 Elena Melikhova, private communication during the 2nd International Scientific Conference on Consequences of the Chernobyl Catastrophe (Geneva, 13 and 14 November 1997).

29 Heinloth, *Die Energiefrage*, p.265.

30 Ibid., p.362.

31 Fritz Vahrenholt, 'Der teure Fortschritt' ('The Expensive Progress'), *Der Spiegel*, No. 52, 1998, p.106.

32 *1998 Britannica Book of the Year*, p.59.

33 Baruch Fischhoff, *Acceptable Risk* (New York: Cambridge University Press, 1982).

34 David Albright, Frans Berkhout and William Walker, *Plutonium and Highly-Enriched Uranium 1996: World Inventories, Capabilities and Politics* (Oxford: Oxford University Press, for SIPRI, 1997); pp.80, 400.

35 Heinloth, *Die Energiefrage*.

Index